Insect Conservation Biology

Michael J. Samways
Invertebrate Conservation Research Centre
Department of Zoology and Entomology
University of Natal, Pietermaritzburg
South Africa

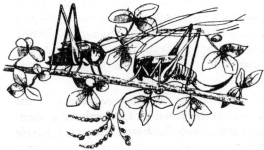

CHAPMAN & HALL

London · Glasgow · Weinheim · New York · Tokyo · Melbourne · Madras

Published by Chapman & Hall, 2-6 Boundary Row, London SE1 8HN, UK

Chapman & Hall, 2-6 Boundary Row, London SE1 8HN, UK

Blackie Academic & Professional, Wester Cleddens Road, Bishopbriggs, Glasgow G64 2NZ, UK

Chapman & Hall GmbH, Pappelallee 3, 69469 Weinheim, Germany

Chapman & Hall USA, One Penn Plaza, 41st Floor, New York, NY10119, USA

Chapman & Hall Japan, ITP - Japan, Kyowa Building, 3F, 2-2-1 Hirakawacho, Chiyoda-ku, Tokyo 102, Japan

Chapman & Hall Australia, Thomas Nelson Australia, 102 Dodds Street, South Melbourne, Victoria 3205, Australia

Chapman & Hall India, R. Seshadri, 32 Second Main Road, CIT East, Madras 600 035, India

First edition 1994
Published in paperback 1995

© 1994 Michael J. Samways

Typeset in 10/12pt Sabon by ROM-Data Corporation, Falmouth, Cornwall
Printed in Great Britain by Clays Ltd, St Ives plc, Bungay, Suffolk

ISBN 0 412 63450 3

A Catalogue record for this book is available from the British Library

Library of Congress Cataloging-in-Publication Data available

Insect Conservation Biology

Conservation Biology

Series Editors

F.B. Goldsmith
Ecology and Conservation Unit, Department of Biology, University College London, Gower Street, London WC1E 6BT, UK.

E. Duffey OBE
Cergne House, Church Street, Wadenhoe, Peterborough PE8 5ST, UK.

The aim of this Series is to provide major summaries of important topics in conservation. The books have the following features:

- original material
- readable and attractive format
- authoritative, comprehensive, thorough and well-referenced
- based on ecological science
- designed for specialists, students and naturalists

In the last twenty years *conservation* has been recognized as one of the most important of all human goals and activities. Since the United Nations Conference on Environment and Development in Rio in June 1992, **biodiversity** has been recognized as a major topic within nature conservation, and each participating country is to prepare its biodiversity strategy. Those scientists preparing these strategies recognise **monitoring** as an essential part of any such strategy. Chapman & Hall have been prominent in publishing key works on monitoring and biodiversity, and with this new Series aim to cover subjects such as conservation management, conservation issues, evaluation of wildlife and biodiversity.

The series contains texts that are scientific and authoritative and present the reader with precise, reliable and succint information. Each volume is scientifically based, fully referenced and attractively illustrated. They are readable and appealing to both advanced students and active members of conservation organizations.

Further books for the Series are currently being commissioned and those wishing to contribute, or wish to know more about the Series, are invited to contact one of the Editors or Chapman & Hall.

Books already published in the Series

For Ben and Camilla, and all
our sons and daughters who will
inherit, care for, restore and
cherish our Earth

Contents

Preface

Why consider conserving insects? If you are empathetic to life on earth, then you have the answer. If you are pragmatic, you may want a more solid answer. In short, they make ecosystems tick. Not only that, they are also numerous, fascinating, varied and economically important. They cannot be ignored.

Insects worldwide are, as far as we can ascertain, disappearing at the rate of thousands of species per year. This is mainly in the tropics where plant diversity and complexity of architecture coupled with thousands of years of climate without severe adversity has allowed millions to evolve and pack themselves into just a tiny fraction of the land surface.

Insect conservation and biotope preservation are inextricably linked, and are integral to biodiversity conservation. Increased insect variety is usually, but not always, associated with increased plant variety. But plant-rich areas may not necessarily be of prime insect conservation concern. Other biotopes may also feature, such as caves or deserts, being home to unique faunas. Species dynamo areas also need conserving, as generators of future diversity.

Prehistory and history have played major roles in determining insect distributions and abundances. Retraction of faunas during the Pleistocene glaciations in the northern hemisphere, retreat to climatic refugia in the tropics and, in modern times, fragmentation of the landscape with increasingly intensive and extensive agriculture and urbanization have all determined present insect population patterns and species distributions.

There are unquestionably different scales of menace facing insect communities. The impacts may range from local river pollution, wetland removal and hedgerow removal to rain forest destruction and global warming. Possibly millions of insect genes are being lost each year, and the trend is escalating.

Entomologists are caught in a dilemma. Some are actively and resourcefully employed in finding ways of suppressing the noxious minority of abundant insects that reduce man's economic returns. Others are aware of the biodiversity crisis and wish to save as many species as possible. Traditionally, applied entomologists have viewed insect conservation with some cynicism, yet all recognize the value of insects in ecological processes. Moot areas exist, such as the realm of insect biocontrol. Yet today there is a meeting of minds on such issues as value and risks of biocontrol, future significance of potentially valuable genetic material and on the important regulatory role played by indigenous predators and parasitoids.

Estimates of the huge number of insect species on earth has promoted awareness of the biodiversity crisis. Insects are extremely important components of the world's biota. This broad introductory text aims to highlight their ecological, economic and intrinsic value within the context of the human-pressurized world. Conservation of genes, species, ecosystems and landscapes in a sustainable way is vital. Let our generation not be viewed as conquistadors plundering resources of enormous potential value. That value is bioempathetic as well as anthropocentrically pragmatic.

This book takes stock of insect conservation biology, the scientific basis for studying and saving as many insect populations, species, their habitats and landscapes as possible. The intention here is not simply to be a story of gloom, but rather to look at practical and bioethical approaches to the whole subject, which is vast, diverse and little researched.

One may question whether there is such a subject as insect conservation biology as an entity, as it is so intimately tied in with landscape ecology, biodiversity conservation and planet management. These may seem big issues, but then insects are a big cog in the biosphere clockwork. Background is important, and so space is given here to prehistorical and historical events prior to overviewing the subject at various scales from the microsite through the landscape to the global perspective. Conservation of the land, and indeed the whole earth, serves the insects' existence. Scale, especially relative to the insects' behaviour patterns, is important and fundamental to asking the right questions leading to appropriate action.

Time is short for conserving the majority of insect species and their ecosystems. Insect conservation biology is fundamental not only to biodiversity conservation, but also to sustainable agriculture and a sustainable biosphere. Broad issues are addressed here. By necessity of space, many issues are only touched upon, with insect examples used to illustrate some of these issues. Pests aside, we need insects, yet we are determining their fate with almost total oblivion. If this book galvanizes action through students, amateur and professional entomologists and, in particular, land and planet managers, then the task will have succeeded. We know insects are important. Do the planners, bankers and politicians know too? This is the direction to go: to emphasize that insect survival is ethical and yet also vital to our well-being.

This book would not have been possible without the interaction and goodwill flow of information from the World Conservation Union (IUCN) and the World Wide Fund for Nature (WWF). The dedication and inspiration of researchers and managers working through the IUCN, WWF, as well as the Xerces Society, the Royal Entomological Society and the Foundation for Research Development (South Africa) stimulated this compilation. Scintillating discussions and correspondence with some of the pioneers of insect conservation biology, including Drs Mike Morris, Tim New, Norman Moore, Jeremy Thomas, Mark Collins, Jeremy Holloway, Jack Dempster,

Bob Pyle, Eric Duffey, Penny Greenslade, Paul Opler, Graeme Ramsay, Frank Howarth, Jan Giliomee, Clarke Scholtz, Nigel Stork and Michael Usher, have focused ideas and emphasized that we really must get moving and relate the significance of insects in ecosystem function to wider conservation circles. Realistic conservationists among many of my students make me realize that their future is our present responsibility.

Very special thanks to Ms Pamela Sweet for processing and reprocessing the manuscript with incredible good humour. Thanks also to Mrs Gael Whiteley, who tracked down copies of papers, and to Mr Billy Boodhoo, who undertook the photographic reproductions, to Ms Rae Osborn for help with checking the text, and to Pepi Ferrari for emphasizing the significance of poetical appreciation of nature. In London, the unwavering confidence in the project by Drs Bob Carling and Clem Earle of Chapman & Hall was a catalyst in bringing the book into being.

Thanks also to the copyright holders, mentioned in the figure captions and reference list, for permission to reproduce illustrations. The illustrations at the beginning of each chapter are from J.G. Wood (1863), and the sketch of two bush crickets on the title page is from Swinton (*c.* 1880), and the endpapers are from Goldsmith (1866.)

The cover illustration, by Professor Denis Brothers, is of a female *Mutilla scabrofoveolata*, a mutillid wasp from South Africa. It represents one of the millions of insects of which we know nothing of their biology, and yet their survival is being increasingly threatened by man's activities.

Every part of this earth is
sacred to my people.
Every shining pine needle,
every sandy shore,
every mist in the dark woods
every clearing and humming insect
is holy in the memory and
experience of my people.
The sap which courses
through the trees
carries the memories of the
red man.
We are part of the earth and
it is part of us.
The perfumed flowers are
our sisters,
the deer, the horse, the great
eagle –
these are our brothers.
The rocky crests, the juices
in the meadows,
the body heat of the pony,
and man –
all belong to the same family.
We know that the white man
does not understand our
ways.
The earth is not his brother,
but his enemy
and when he has conquered
it he moves on.
He treats his mother
The Earth
and his brother
the sky
as things to be bought,
plundered,
sold like sheep or bright beads.
His appetite will devour the
earth
and leave behind only a
desert.
The sight of your cities
pains the eyes of the red man.
But perhaps it is because
the red man is a savage
and does not understand.
There is no quiet place
in the white man's cities.
No place to hear
the unfurling leaves in
spring
or the rustle of an insect's
wings.
What is there to life
if a man cannot hear

the lovely cry of the
whippoorwill
or the arguments of the frogs
around a pond at night?
The Indian prefers
the soft sound of the wind
darting over the faces of a
pond
scented with the piñon pine.
The air is precious to the red
man,
for all things share the same
breath - the beast, the tree,
the man.
The white man does not seem
to notice the air he breathes.
The air shares its spirit with
all the life it supports.
I have seen
a thousand rotting buffaloes
on the prairie
left by the white man
who shot them from a passing
train.
I am a savage
and I do not understand
how the smoking iron horse
can be more important than
the buffalo
that we kill only to stay alive.
You must teach your
children
that the ground beneath
their feet
is the ashes
of our grandfathers.
Tell your children
that the earth is rich
with the lives of our kin.
Teach your children what we
have taught our children
that the earth
is our mother;
whatever befalls the earth
befalls the sons of the earth.
If men spit upon the ground
they spit upon themselves.

This extract, although said to be from a letter from Chief Seattle to Franklin Pearce, President of the United States of America, 1855, is in fact a hoax. It was written in 1971/72 by Ted Perry for the film *Home* (see Hargrove, E. (1989) The Gospel of Chief Seattle is a hoax. *Environmental Ethics*, 11, 195–196). Although not authentic, the message is nevertheless poignant.

Part One
Setting the Scene

Global variation in insect variety

Scientists in their quest for certitude and proof tend to reject the marvellous.

Jacques Cousteau

What ... is the mean diameter of the Earth? 12,742 km. How many stars are there in the Milky Way? 10^{11}. How many genes in a small virus particle? 10 ... What is the mass of an electron? 9.1×10^{-28} grams. How many species of organisms are there on Earth? We don't know, not even to the nearest order of magnitude.

Edward O. Wilson

1.1 INSECT SUCCESS

1.1.1 A variety of biotopes

While the metallic blue glint of a morpho butterfly catches the eye of a naturalist in the South American forest, a mosquito surreptitiously sucks blood from his leg and an ant wriggles helplessly under his boot. A colleague high in the North American Rocky Mountains notices a boreid snow scorpionfly moving through the inhospitable landscape in search of a meal of moss. Insects are virtually everywhere on the earth's surface, excluded only by the extremes of climate at the poles and on the peaks of the highest mountains: just a few species live in the sea (Cheng, 1976).

Beetles of the genus *Helophorus* occur in the hot springs of the Himalayas at 5400 m above sea level, and some anthomyiid flies eke out a living 6200 m above sea level in the Mount Everest region (Mani, 1968). Springtails (Collembola), which are insects only in the very widest taxonomic sense, are common in some parts of Antarctica. Tenebrionid beetles survive in the searing heat of the Namib Desert by feeding on detritus caught up in miniature wind vortices around the dunes and collecting moisture for drinking from droplets of the early-morning mist (Louw and Seely, 1982). A rhaphidophorid cricket may scavenge in the permanent quiet and darkness of a cave, while a sessile, adult, female red scale (*Aonidiella aurantii*) suffers daily a temperature of 60°C in the sun on the surface of a citrus leaf. The fly *Psilopa petrolei* has adapted to living in pools of crude petroleum, while lice have been recorded from the bodies of many birds and mammals. In the process of adapting to living in such a vast variety of biotopes, insects have also characteristically been able to feed on a variety of resources. These include seeds, leaves, flowers, bark, hair, feathers, vertebrate blood, bone and the ash of burnt grass. About 15% are parasitic on other insects; in some cases adelphoparasitism has arisen, as in the wasp *Coccophagus utilis*, where the male parasitizes the female.

1.1.2 Structure and mode of life

The success of insects is partly due to their highly effective external skeletal structure, which is lightweight, tough and waterproof. But it has limited their size: an insect the size of a tortoise would implode under its own muscular exertion.

Nevertheless, small size has enabled insects to inhabit the vast array of crevices and interstices in the complex architecture of plants, the soil, leaf litter and the bodies of other animals, including other insects.

The insect's light and strong exoskeleton has given it flight, sustained by an effective tubular tracheal system which conveys oxygen to the muscles

800 000 times faster than if the gas had to diffuse through the tissues (Price, 1984). Flight has permitted efficient searching for distant resources that otherwise by walking would be too far away to exploit. Flight also means shorter time needed for finding a mate, especially when coupled with mate-attracting pheromones. The wing has also given migration, and the opportunity not only for finding more valuable food resources, but also for dispersing eggs in areas where the offspring might better benefit.

Another feature of the insect exoskeleton is its adaptiveness to a myriad of forms and colours. The survival advantages of mimicry in both form and colour of plant parts and of other insects has driven diversification in many insect groups.

Despite the estimated several thousands of genes in each insect, possibly up to 80% of the DNA is not encoded. Yet there is sufficient variety in the remaining 20% not only to produce the vast variety of insect species but also many polymorphisms besides mimics. Seasonal forms are not uncommon in some butterflies, where the dry-season form may appear as being quite different from the wet-season one. Sexual dimorphism is also sometimes so extreme that the male, from a functional ecological viewpoint, may be quite a different animal from the female. Such is the case with diaspid scales. The female is a sessile, long-lived armoured disc, looking like a spot on a plant leaf, with its 1-cm-long fine hair-like feeding proboscis coiling into the plant tissue. The male, in contrast, is a tiny airborne form, living only a few hours, serving little other purpose than to find a mate to fertilize (Figure 1.1). From a conservation perspective, environmental conditions must consider the various morphs, behavioural patterns such as migration and the requirements of both sexes.

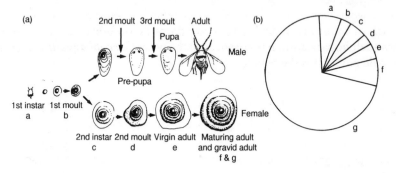

Figure 1.1 (a) Developmental stages of the diaspid scale, *Aonidiella aurantii*, leading to extreme sexual dimorphism in the adults. (b) Duration of each stage at 26°C. Total = 120 days, although the male lives only a few hours and the female many weeks. In reality, the female is many hundreds of times the volume of the male. Conservation of insects must consider the possibility of quite different ways of life and different biotopes of the two sexes. (From Samways, 1988a.)

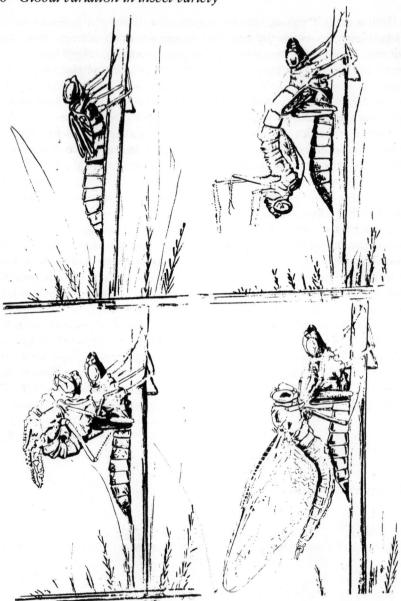

Figure 1.2 Developmental polymorphism in some exopterygote insects (such as the *Aeshna* sp. dragonfly here) and many endopterygotes can mean that, from a functional ecological point of view, the young stages are different organisms from the adults. Here the larva has left the medium of the water to emerge into a highly aerial adult. In an insect such as this dragonfly, both the larva and the adult are predatory, albeit in different media. For an endopterygote such as a butterfly, the medium may be the same, yet the larva is a herbivore, while the adult is nectar feeding. (From Hammond, 1983.)

Of even greater significance in conservation biology is developmental polymorphism, in which the immature stages are quite different from the adult not only in form, but also in feeding and many aspects of behaviour. Survival of the species depends naturally on maximal or optimal survival of all the developmental stages. This co-survival is a linear phenomenon, as illustrated in life tables, with each successive stage dependent on the number of survivors from the previous stage. This is particularly important in the endopterygotes, such as the butterflies: suitable conditions for the folivorous, relatively immobile larva are quite different from the nectar-feeding, aerial, highly mobile adult. Such special requirements of the different life stages have, in evolutionary terms, led to great species diversity in some groups such as lycaenid butterflies. In turn, it has made them particularly vulnerable to anthropogenic impacts.

Developmental polymorphism is also an important consideration in the conservation biology of some exopterygotes, as some are archaic and unique in form and distribution and therefore of genetic curiosity. Often the larvae live in a different medium from that of the adults (Figure 1.2). Important groups are the Ephemeroptera, Odonata and Plecoptera, several of which are listed in *The IUCN Invertebrate Red Data Book* (Wells, Pyle and Collins, 1983). The larvae are mostly aquatic, whereas the adults are aerial, both media posing quite different intensities or types of threat from anthropogenic disturbance. The relict Himalayan dragonfly (*Epiophlebia laidlawi*) is one of two surviving species of the suborder Anisozygoptera which flourished in the Mesozoic Era. This order combines features of the damselflies (Zygoptera) and the dragonflies (Anisoptera) and is of great evolutionary interest. Deforestation, reclamation for cattle grazing and tourism are directly modifying the adult biotope, and leading to soil erosion which is silting the clear streams of the larva.

1.1.3 Insects on plants

An extraordinary paradox in the insect world is that although cellulose is the most widespread of all terrestrial biological substances and the insects-feeding-on-plant tissue is the most common of all interspecies biotic relationships, few insects apparently possess the enzyme cellulase. Apart from some Thysanura, Coleoptera and termitid termites, most others that digest cellulose do so either using gut symbionts that produce cellulase, such as some bacteria and protozoa, or they cultivate fungi to break down the cellulose. In the majority of plant-feeding insects, the cellulose is either passed through the gastrointestinal tract unutilized, or the insects avoid the problem by siphoning plant liquids through a proboscis which penetrates the plant tissue.

True plant-feeding insects make up about a quarter of all living species, even excluding the multitudes of insects that feed on algae, bore into wood or feed on leaf litter. Despite the obstacles posed by desiccation, attachment, nourishment and plant chemical defences, over half of the insect orders today are phytophagous (Southwood, 1973; Strong, Lawton and Southwood, 1984). The conservation implications are that plant preservation soundly underpins many aspects of insect conservation.

Insects are also important as pollinators of flowering plants, in both natural and agricultural ecosystems. The angiosperms originated about the Jurassic–Cretaceous boundary (135 million years ago), and became the dominant plant group. Although some groups such as the Lepidoptera flourished and radiated, other groups continued to feed on more primitive plant groups or disappeared. Survival of both plant and pollinator depends on conservation programmes that preserve both mutualists. The Comet orchid (*Angraecum sesquipedale*) of Madagascar has a tubular nectary which is 30 cm long. Alfred Russel Wallace in the last century predicted that it must be fertilized by a moth with a proboscis 30 cm long. Decades later the moth, a hawkmoth, *Xanthopan morgani praedicta,* was discovered. Interestingly another Madagascan orchid, *Angraecum 'longicalcar'* has been discovered which has a nectary spur 36–41 cm long. Its pollinator, presumably another hawkmoth, has not yet been discovered (Schatz, 1992). Conservation of both the moths and the plants are intimately related, and conditions that support not only the mutualism but also the food plant of the larva and the seeds of the plant must prevail. Such a situation becomes increasingly acute for widely dispersed, rare plants with low apparency and special pollination and biotope requirements. It is additionally important in the preservation of unique and ancient taxonomically remote plant and insect taxa. We do not know the extent to which specific plants depend on particular insects for pollination, or the degree to which certain insects depend on specific plants. Some of the ancient plant groups may in fact be ecologically tolerant with regard to insects, as illustrated by their continued long existence despite changes and other abiotic and biotic stresses over time. In reality, it may be the modern finely tuned mutualisms between recent angiosperms and recent insect species that are the least robust and in greatest need of protection.

1.2 INSECTS IN ECOSYSTEMS

1.2.1 Numbers of individuals, energy flow and biomass consumption

Stork (1988) calculated the massive figure of over 42 million arthropod individuals per hectare for Seram rain forest, living in the vast number of microsites in the soil, leaf litter, ground vegetation, tree trunks and canopies (Figure1.3). The soil contributed most, dominated by Collembola and mites. Wilson (1991) estimated that, for the 9000 or so living ant species, at any one

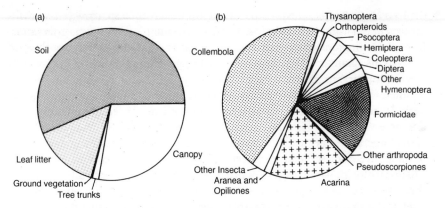

Figure 1.3 Relative proportions of the estimated 42.3 million arthropods in a hectare of Seram rain forest (calculated from data resulting from soil, leaf litter, ground vegetation, tree trunk and canopy samples). (a) The contributions of the five major biotopes. (b) Contributors of the major arthropod groups. (From Stork, 1988.)

moment in time there are 10^{15} living individuals of just this one family.

Insects and other arthropods are abundant throughout terrestrial ecosystems, probably removing about 20% of the foliage annually worldwide. But insects do not function in isolation, particularly in the soil, where in association with many other detritivore invertebrates, fungi and bacteria, they release and recycle the nutrients locked up in decaying vegetation. This turnover is especially rapid in the tropics, which are essentially deserts covered with trees dependent on this rapid recycling of nutrients.

It is not particularly ecologically interesting to know exactly how many individuals occur in one area at one time, as there is so much variability with microclimate and season. The point here is that insects are numerous, and their conservation biology is intimately linked with ecosystem and global survival.

There is enormous variation in biomass from one taxon to another, and between areas. In the tropical forests, ants are particularly numerous. Near Manaus, Brazil, ants and termites make up more than one-quarter of the faunal biomass, and ants alone have four times the biomass of all land vertebrates combined (Wilson, 1991). These insects, together with the social wasps and bees, constitute 80% of the biomass. Ants scavenge 90% of the dead remains of all insects and are a major element in the turnover of soil (Wilson, 1991).

Odum, Connell and Davenport (1962) illustrated the importance of insects in the herbivore community in old fields in South Carolina. The foliage-eating grasshoppers *Melanoplus femur-rubrum, M. biliteratus* and the tree cricket *Oecanthus nigricornis* together showed an average energy flow per year of

25.6 kcalm^{-2}, compared with 3.6 kcalm^{-2} for the savanna sparrow (*Passerculus sandwichensis*), and 6.7 kcalm^{-2} per year for the mouse *Peromyscus polionotus*. Indeed, Orthoptera alone can have a major impact on the energy flow of ecosystems. Gandar (1982) found grasshopper biomass during one year in the South African savanna to be 0.73 kgha^{-1}. At this density, the insects ingested 94 kgha^{-1} of plant material per year. In addition, during the same period, an extra 36 kgha^{-1} was wasted in the general feeding process. The outcome was that grasshoppers removed up to 16% of the grass cover.

As insects can play such a major role in ecosystem processes, conservation of land areas may need to determine the role of insects in that area. Theoretically, conservation of a pristine landscape automatically conserves the insects and the ecological processes in which they are involved. But as undisturbed areas are now virtually non-existent, and those that do exist have disturbed edges, the disturbances will inevitably tip the balance in favour of presence or absence of species, or will affect their relative abundances. Ecological processes will therefore change, and where an insect is an ecologically keystone species its numbers may increase to reach outbreak numbers. The tingid bug *Vatiga illudens* is relatively rare in the Brazilian savanna, but reaches pest proportions on cassava when planted in the area (Samways, 1979).

1.2.2 Limiting factors on insect abundance

Varley, Gradwell and Hassell (1973) have outlined factors that reduce insect population levels. Abiotic factors such as climate and weather reduce populations irrespective of the density of individuals, i.e. they are density-independent factors. Competition between individuals for limited resources, albeit variable in nature, increases with population density, i.e. density dependence operates. Competition between species may eliminate one, the competition increasing with increase of the winning species, or both species may continue to coexist, i.e. density-dependent and non-linear density-dependent factors are operating respectively. Parasitoids, predators and pathogens are usually density-dependent factors. But this effect may vary with time, with physical quality and spatial characteristics of the biotope, and with changes in the species composition of the community.

All these factors play a role in determining the survival of local populations. The impact of weather (section 6.4) is known to drastically reduce some insect populations (Figure 1.4). This is likely to be an increasingly significant mortality factor with global climate warming and variability in weather conditions. Also, with reduction in available biotopes as wildlands are cultivated or built upon, resources become shorter with concurrent increasing intraspecific density-dependent competition. This is a real threat, as it is already known that some insect populations are held down by limited resource availability (Dempster and Pollard, 1986; Pollard, 1981).

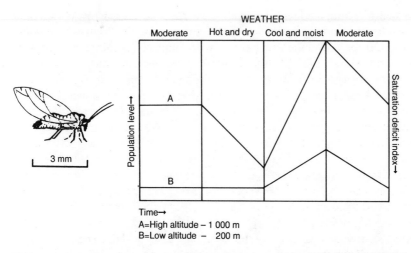

Figure 1.4 Schematic summary of the relationship between adult citrus psylla (*Trioza erytreae*) population levels and weather at two different altitudes in southern Africa. The saturation deficit index (SDI) is a measure of climatic severity, and equals saturation vapour pressure (SVP) (at maximum temperature) minus minimum relative humidity multiplied by SVP (at maximum temperature). (From Samways, 1987.)

Anthropogenic modification of the landscape inevitably shifts the competitive advantage of one species over another. This may not necessarily be direct, but may be indirect through changes in plant composition. J.A. Thomas (1991) has emphasized that the young stages of many British butterflies are more sensitive to microclimatic change than was previously thought. These requirements have restricted the larvae to narrow niches within their biotopes, usually corresponding to a short-lived seral stage and often to a warm microclimate. Land use changes can readily change microclimatic conditions, generally adversely for most species, although sometimes benefiting a few.

Parasitoids and predators are well known to restrict population growth (Hassell, 1985), sometimes acting differentially depending on the population level of the host (Samways, 1988a). Insect conservation biology implies that these parasitoids and predators are also subjects themselves, with loss of a host impacting upon its dependent community, mutualists as well as parasitoids and predators. Predation from vertebrates is also an important consideration, hence the evolution of aposematic colouring (Guilford, 1990) and mimicry (Malcolm, 1990). Translocated and introduced insect predators and parasitoids are the regular subjects of successful classical biological control (Caltagirone, 1981), while introduced vertebrate predators have had some catastrophic impacts upon insect populations (Chapter 8).

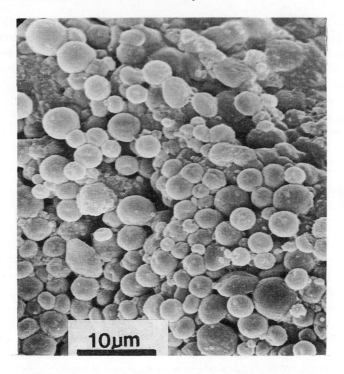

Figure 1.5 The high density of spores of the pathogenic protozoan *Nosema* sp. in the faeces of the moth *Pareuchaetes aurata*. (Photograph by kind courtesy of R.L. Kluge and P.M. Caldwell.)

Some insect species are vectors of plant disease, the causal pathogens being mostly viruses, fungi and bacteria. Such diseases can affect the competitive advantage of certain plants, which has an indirect effect upon insect conservation by altering microhabitat and even biotope.

Insects themselves are subject to a wide range of pathogens (Figure 1.5), including viruses, rickettsiae, bacteria, protozoa, fungi and nematodes. Degree and type of infestation are often affected by prevailing weather conditions. Mortality may vary from one population, and one species, to another. Although only 4.5% of fully grown larvae of the winter moth (*Operophtera brumata*) in Britain die from microsporidian protozoa infection (Varley, Gradwell and Hassell, 1973), local populations of various Homoptera can be entirely eliminated by the fungus *Cladosporium oxysporum* (Samways and Grech, 1986).

For insect conservation biology these limiting factors on insect abundance are immensely varied and complex, even for one species in one area at one time. These various and varying mortality factors, when set against the high natality characteristic of so many insect species, together are responsible for the great population variation and even complete crashes seen so often in the insect world. This suggests that finding requirements necessary for the survival of a single species, although appropriate for certain high-profile species, is, for most insects, simply not all-embracing enough for realistic conservation of so many insect species. Landscape conservation and appropriate management is a more fruitful approach.

1.3 WORLD INSECT SPECIES RICHNESS

1.3.1 Numbers of described species and the species concept in conservation biology

Hammond (1992) estimated there to be 950 000 described insect species, although lower figures of around 750 000 and 790 000 (May , 1990a) are generally quoted. The larger figure represents 56.4% of all described biota according to Hammond (1992), and almost 77% of all metazoan animals according to various sources (Table 1.1). Were all species of insects described, this figure would rise to well over 90%. Many of these are likely to be synonyms, while others are yet-to-be-separated sibling species, morphologically indistinguishable. Further, there is the debate as to what exactly

Table 1.1 Approximate number of described species on earth

Viruses	1 000
Monera	5 000
Fungi	47 000
Algae	27 000
Plantae	250 000
Protozoa	31 000
Porifera	5 000
Coelenterata	9 000
Platyhelminthes	12 000
Nematoda	12 000
Annelida	12 000
Mollusca	50 000
Echinodermata	6 000
Insecta	950 000
Non-insect arthopoda	125 000
Minor invertebrate taxa	10 000
Chordata	44 000
Total	1 596 000

Note: A distillation from many sources.

constitutes a species (Adis, 1990; May, 1990a,b).

The phylogenetic species concept of Nixon and Wheeler (1990), which is the smallest aggregation of populations (sexual) or lineages (asexual) diagnosable by a unique combination of character states in comparable individuals (semaphoronts), is sufficiently tangible and robust for conservation biology purposes. Species concepts based on implications of ancestry are difficult for practical management. Even the invoking of genetic concepts poses problems in view of the inferential findings of Houck *et al.* (1991) that the semiparasitic mite *Proctolaelaps regalis* can transfer genes from one *Drosophila* species to another.

1.3.2 The tropical rain forest established as the hub for insect species richness

Erwin (1982), in a study of the specificity between beetle and tree species in the Panamanian rain forest, estimated that instead of there being 1.5–3.0 million species (described and undescribed) on earth, as was previously thought, there may be as many as 30 million species of arthropods (principally insects) alone. With further information, he has increased his estimate to 50 million species (Erwin, 1988).

Stork (1988) reappraised Erwin's (1982) data and emphasized difficulties in estimating insect numbers based on host specificity, and also pointed out that some assumptions may be incorrect, e.g. that beetles represent 40% of the canopy arthropod species. The reanalysis by Stork (1988) with additional data from South-East Asian rain forests, produced the interesting suggestion that 70% of arthropod species occur in the leaf litter and 14% in the canopy. But still the outcome was that there may be between 10 and 80 million species on earth.

May (1986, 1988, 1989) has tackled the question of how many species on earth from a different viewpoint (Figure 1.6). The log–log plot of the number of species as a function of the physical size of all individuals gives an almost linear regression line, with a slope close to –2, for species with body lengths upward of about 1 cm. This means that for the relatively well-known larger species, a 10-fold reduction in linear dimensions results in roughly a 100-fold increase in species. However, with the smallest animals our knowledge, and the even size–number relationship, becomes increasingly uncertain. Nevertheless, extrapolating down to a body length of about 0.5 mm, May (1989) suggests that there are about 10 million animal species.

Gaston (1991) approached the problem subjectively, basing estimates on the knowledge of expert taxonomists, and has suggested that a figure of 10 million species of insects is tenable and one of around 5 million is feasible. Meanwhile, Hodkinson and Casson (1991), using sampling methods in the tropical rain forests of Sulawesi Utara, deduced by extrapolation that the

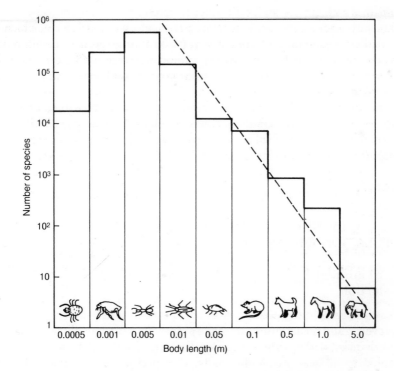

Figure 1.6 An approximation of the numbers of all described (the histogram) terrestrial animals categorized according to body length. The total number of species, undescribed as well as described, would change the shape of the left-hand side of the histogram, and increase the height of the bars. A much greater proportion of the larger animals have been described, with the right-hand side of the histogram having a characteristic slope. (From May, 1989.)

world total of insects is between 1.84 and 2.57 million species. Similarly, Gaston (1992), using estimates based on insect/plant species ratios, also suggested a lower figure, somewhere between 2.75 million and 8.75 million species of insects.

Erwin's (1982) pioneering work has highlighted that there are many species on earth, but we are a long way from actually confirming the number. The favoured average figure is somewhere just under 10 million. For the purposes of inventorying global biodiversity, a working figure of 8 million species of insect and 8.9 million species of arthropods is currently used, with a world total of all 'species' (including viruses and bacteria) being about 12.25 million (Hammond, 1992).

Briggs (1991) has entered the debate, estimating that 98% of all animals live on land, just 29% of the earth's surface. This figure is based on an estimate that there are about 12 million terrestrial metazoan species, compared with a total of 160 000 marine species. Grassle and Maciolek (1992) have since argued that the estimate of 160 000 marine species is certainly too low, and the sea may hold between 1 million and over 10 million species. If this is the case, then terrestrial species may account for between about 50 and 90%, rather than 98%.

Only between about 5 and 10% of insects have scientific names. Of those, there is little knowledge on their genetics, behaviour, ecology, distribution, abundance or economic value. In terms of insect conservation, the pressing problem is not determining in fine detail the actual number of species, but how to conserve as much and as soon as possible of this enormous genetic variety.

1.4 LATITUDINAL GRADIENTS IN SPECIES RICHNESS AND POPULATION VARIABILITY

1.4.1 Latitudinal gradients in climate, weather and primary production

Organisms in cold-temperate areas are exposed at different periods of their life cycles and generations to highly varying seasonal climate changes (Dobzhansky, 1950). There is natural selection for organisms to be well adapted to all seasons. Additionally, the species must be tolerably well adapted to occasional extremes of seasons. Such changeable environments put the highest premium on versatility rather than on adaptive perfection.

Species are not adapted to average conditions, but adverse ones. This is highly relevant to ectotherms such as insects in extremely cold years in temperate lands. They have responded adaptively with diapause. Adverse conditions occasionally occur in the tropics, where water droplets from storms and dew entrap small insects through surface tension effects (Samways, 1979).

Many animal assemblages increase in species richness towards the tropics (Fischer, 1960; Pianka, 1966; Schoener and Janzen, 1968; Rapoport, 1982). Additionally, Stevens (1989) has drawn attention to 'Rapoport's rule', that many species have narrower latitudinal geographical ranges at lower tropical latitudes, i.e. species occur in narrower bands at the equator than in temperate areas. Indeed, it is in the tropics where insect species packing reaches a maximum, with an estimated 55% of all insects living in tropical rain forests, less than 7% of the world's land surface.

In some insect groups, such as the termites, the trend in both species richness and biomass reaches extremes, with very high biomass and numbers in the tropics, yet absence from the highest latitudes (Figure 1.7). The latitudinal species gradient is fairly distinctive in some insect groups, such as

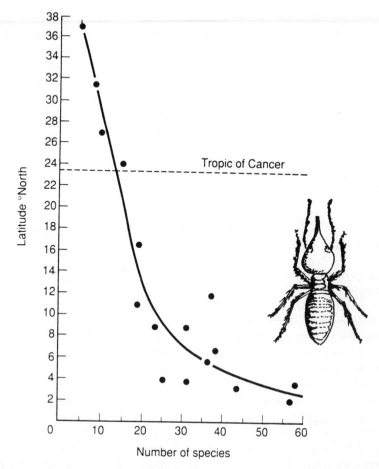

Figure 1.7 The number of termite species according to latitude, north of the equator. (Graph from Sutton and Collins, 1991.)

the swallowtail butterflies (Collins and Morris, 1985), ants (Kusnezov, 1957) and stream insects (Stout and Van der Meer, 1975). The rule, however, is not general for all insect groups. Parasitic wasps (Owen and Owen, 1974) and Collembola (Rapoport, 1982), show the highest species richness in mid-temperate regions. Excluding the extreme polar regions, aphids and psyllids tend to show a reverse gradient trend (Eastop, 1978). In psyllids, this is in part because the young stages are sensitive to stressful conditions of high temperature and low humidity (Samways, 1987), which are prevalent in the subtropics. This may not apply to the psyllid fauna of the equator, where conditions are moist and relatively stable. It may simply be that insufficient collecting has been done in the tropical zone.

Primary production does not appear to be a driver for these latitudinal gradients. Walter (1985) has calculated that beech forests in central Europe are no less productive than equatorial forests (13.5 tha^{-1} and 13.4 tha^{-1} annually respectively), and Whittaker (1975) gives production values only 50% higher for tropical than temperate forests, and temperate grassland is only marginally less productive than subtropical savanna. Within the tropics, increased plant species richness may not necessarily be correlated with high productivity (Tilman, 1982), but nevertheless generates high insect species richness.

1.4.2 Precipitation, climate and plant variety

Increased precipitation usually increases insect species richness. Exceptions are the lowland dry rain forest and an adjacent lowland rain forest in northern Costa Rica, which have similar species richnesses among the Lepidoptera and Hymenoptera, with a species overlap as high as 80% for sphingid moths (Janzen , 1988). For highly stenotopic groups however, this figure is likely to be much lower in the dry areas than in the wet ones.

Arid areas, to the north and south of the humid tropics, are relatively species poor, although the species do not necessarily show less endemism or interesting behavioural attributes than in the wet tropics. It is simply that plant communities, and therefore plant biochemistry and architecture, do not generate the same number of niches as in the tropical forest. Also, physical environmental conditions are more extreme diurnally, seasonally and annually in the desert and savanna than in the tropics. This results in species that tolerate frequently varying, adverse conditions. The crucial factor is not whether conditions are stressful *per se*, but whether they are predictable or not. High stress with a high degree of predictability of the type, pattern and degree of stress generates high species richness, but stress that changes randomly in type, pattern and/or degree tends to reduce species richness as the time comes sooner or later when certain species cannot survive the harshest conditions.

J.H. Brown and Davidson (1977) found that, among North American seed-eating ants, species richness increases with precipitation and primary productivity independent of latitude. Yet in contrast, the semiarid area of southern Australia has an ant species density of up to 150 ha^{-1}, similar to that of the density of the lowland equatorial forests (Greenslade and New, 1991). In Australia, the trend is for ant species diversity to decrease towards the southern latitudes of the continent (Figure 1.8). The contrasting North American and Australian findings strongly suggest caution when extrapolating from one ecological area to another. The biogeographical data, in turn, warn that distributional trends can also vary from one taxon to another. In conservation terms, this means each geographical area needs at least inventorying to determine which species are present. Following on from this is then

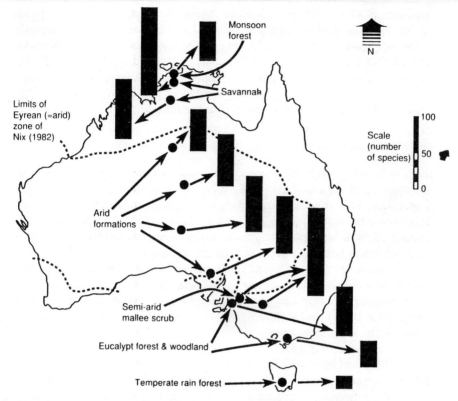

Figure 1.8 Schematic map of numbers of species of ants in different vegetation types in a transect across Australia. [From Greenslade and New (1991), based on published and unpublished data of P.J.M. Greenslade.]

an assessment of the value of that biota.

1.4.3 Possible factors for high insect richness in the tropics

The debate on why there is so much biotic diversity in the tropics is far from resolved. Lack of climatic harshness over millions of years has been a likely influence directly on plant and insect form and physiology, with both groups in turn influencing each other.

It seems that plant chemistry, the coevolutionary interaction between plant and insect and the role of novel defensive and counter-defensive characteristics seem to have contributed to the diversification of insects (Mitter, Farrell and Futuyma, 1991). But such diversification is restricted by taxonomic categories. Evidence from the strong phylogenetic component in many insect–plant associations suggests that insects are not able to readily change hosts. In many cases, the insects' diet may not be optimal owing to these constraints of evolutionary history.

The great insect variety in the tropics may be due not only to climate, plant taxonomy and physiology, but also to plant architecture and plant allopatry. Both plants and insects continually form small populations within the climatic constancy of the area. This readily leads to isolated populations, high survival rates and great speciation among the isolated population pockets. In the warm conditions, where multivoltinism is commonplace, there is rapid allopatric speciation between one small area and another.

The climatic constancy of the tropics is often punctuated by short-term and/or localized disturbance. This may be from tree fall, river fluctuation or an exceptionally heavy downpour among many other possibilities. These intermediate level disturbances, which have continued over millenia, have generated high species richness, with some insects' survival depending, for example, on inevitable tree-fall gap disturbance (Braker, 1991).

In addition, folivorous insect species diversity is enhanced by predation and parasitism (Lawton and Strong, 1981). Increased herbivore richness in turn generates increased parasitoid, hyperparasitoid, predator and mutualist richness. In reality, it is probably the combination of all these factors that has variously promoted insect diversity.

1.4.4 Latitudinal gradients in species population variability

The tropics although relatively climatically constant, can have distinct daily and seasonal changes in weather (Wolda, 1988). Where rainfall shows a steady annual pattern, there may still be variations in the distribution of diurnal rainfall and daylight hours throughout the year.

Measurement of population variability is fraught with analytical difficulties, depending on whether absolute or proportional variability is being measured (Samways, 1990a). This conclusion was based on a study of subtropical ant assemblages, and indicated that common species are not more variable than rare ones, but the rare species tend to be stenotopic and always rare, whereas the common species are eurytopic, highly competitive and demographically volatile.

In general however, tropical and temperate species frequently show great variations in seasonal abundance (Dobzhansky and Pavan, 1950; Wolda, 1988). In the context of insect conservation biology, this means that population variation monitoring is just as important in the tropics as in temperate lands, where some long-term methods have already been developed (e.g. Pollard, 1982). The difficulty is that there is not the manpower to do this even for a fraction of the endangered tropical species. Landscape conservation and preservation, monitoring of indicator groups, such as butterflies, dragonflies (Sutton and Collins, 1991) or cicindelid beetles (Pearson and Cassola, 1992), is the most important and urgent umbrella option at the present time.

1.5 INSECT AND PLANT DIVERSITY TRACKING: NORTHERN AND SOUTHERN HEMISPHERES

1.5.1 Comparisons of the African Cape with areas to the north

In the southern hemisphere, as at the equator, there are nutrient-poor soils that have escaped Pleistocene glaciations and have today an extraordinarily rich plant fauna. The Cape floral kingdom, although only 89 000 km^2 in extent, has an extremely rich flora of 8550 species in 957 genera, of which 73% of the species and 21% of the genera are endemic. About 68% of these are threatened (Hall *et al.*, 1984). The change of species composition from one patch to another is high, with two sides of the same valley differing by 45%, and adjacent large areas sharing less than 40% out of more than 2550 species. The Cape Peninsula, only 470 km^2 in area, supports 2256 indigenous species, more than half the flora of eastern North America (Huntley, 1988).

The Cape floral kingdom is also extremely rich in localized endemic insect species, including many lycaenid butterflies which have very restricted ranges and are threatened by local landscape changes and by global events (Samways, 1993). However, it does not seem to be the plant taxonomic variety *per se* that has generated the insect diversity, but that the insects have speciated through allopatric isolation as did the plants themselves. Also, there has been considerable conservatism on the part of the insects in maintaining coevolutionary relationships with particular plant taxa rather than radiating with the plant species (Cottrell, 1985).

Interestingly, another heathland, the managed *Calluna* landscape of northern England, also illustrates that increased plant species richness does not necessarily increase arthropod richness (Usher, 1992). However, the difference between the Cape and northern England is the vast number of endemic species in the Cape that have probably been generated allopatrically through lack of glaciation and intricate topographic patterns over tens of millions of years. England's fauna is post-glacial and invasive and the interactions more dependent on landscape patterns (Webb, 1989) than on ancient events.

A further interesting comparison is the Cape with the tropics. Neither was sterilized by Pleistocene glaciations. This absence of mortally cold conditions was an enabling process, providing sufficient time for a variety of species to be able to evolve. Overlying this long time scale and large spatial scale of events, have been the local events. Just as the bushy plants on a complex topography have stimulated allopatric insect speciation in the Cape, so the complex treescape on a flatter surface has probably done for the tropical insects.

1.5.2 Swallowtail butterfly trends

There are about 573 species of swallowtail butterfly (Papilionidae) world-wide, with by far the majority occurring in the tropics (Collins and Morris, 1985). Any comparisions of species richness between the northern and southern hemispheres must, at this stage in our knowledge, assume that latitudinal ranges, and geographical range compression towards the tropics (i.e. Rappoport's Rule (Stevens, 1989)), are similar in the two masses. Although the north has two-thirds of the earth's land mass, up to 30° latitude either side of the equator the land masses are fairly equal in area. This means that at least for the tropical and sub-tropical areas the north and south are very similar in swallowtail species richness. Proportionate to the shrinking land areas in the southern temperate area this similarity in trends beyond north and south continues up to about 40°. Beyond 40°, the continental north is considerably richer than the land-poor south at the same latitude. Naturally the cold biomes of the far north, however, are species-poor compared with the tropics

1.5.3 North *vs.* south: insect species per area and per plant species

This north versus south trend is not necessarily consistent for all insect taxa combined, whether for more insect species per plant, or for more insect species per unit area of land (Table 1.2).

The USA and Australia have similar ratios and similar numbers of insects per area. In contrast, Britain has well over three times as many insects per plant species as the USA or Australia. Southern Africa has twice as many as these two countries. Southern Africa is insect species-rich per unit area, but closely followed by Britain.

Although these figures are approximate, they do nevertheless suggest that the southern hemisphere is not necessarily richer in insect species than the northern hemisphere. The surprisingly rich fauna of Britain is made up of post-glacial invasive plants which have carried with them a heavy insect species load. Globally it appears that in temperate and Mediterranean–sub-tropical climatic areas, insect diversity does not track plant diversity at the same rate. Similarly, insect species richness per unit area can be extremely high whether in the northern hemisphere (Britain) or in the southern hemisphere (southern Africa). Although evidence from different insect groups varies, rough estimates of the total number of insect species in different countries suggest that there may be a tendency for insect-to-plant species ratios to be proportionately lower when plant richness per unit area is high (Gaston, 1992).

There is, in insect conservation biology terms, a vitally important rider. Percentage levels of endemism can be high in southern Africa [e.g. Apoidea 85% (Eardley, 1989); Tettigonioidea about 75% (Ragge, 1974)]. Even taxa

Table 1.2 Approximate ratios of numbers of insect species to numbers of higher plant species, and numbers of insect species per 1000 km^2, in some northern hemisphere compared with southern hemisphere biogeographical units

Biogeographical unit	Area (1000 km^2)	No. of insect species	No. of plant species	Ratio	No. of insect species per 1000 km^2
Britain	245	22 400[1]	1 500[2]	14.9:1	91.4
USA	9363	89 000[3]	20 000[4]	4.45:1	9.5
Southern Africa	2573	250 000[5]	22 977[6]	10.9:1	97.2
Australia	7686	125 000[7]	25 000[4]	5.00:1	16.3

Sources: [1] Fry and Lonsdale (1991); [2]Turrill (1948); [3]approximate estimate and extrapolation from Borror and White (1970); [4]Durrell (1986); [5]approximate estimate from 80 000 described species (Prinsloo, 1989); [6]Gibbs Russell (1985); [7]probably a conservative estimate (Greenslade and New, 1991).

that tend to show low percentages of endemism in most places of the world can be fairly high in southern Africa. Acridoidea are estimated to be 47% endemic to South Africa (Johnsen, 1987), the highest percentage value for any country on the continent. Even the Odonata, with their high mobility, are 18% endemic to South Africa (Samways, 1992a). Compare these figures with the mere 10 endemic British taxa out of the 13 746 listed species of Shirt (1987).

This cautions against making conservation decisions on the basis of general global trends. It indicates the importance of recognizing the taxonomic or other characters of particular biotopes, ecosystems, landscapes and biomes around the world. Islands are often such special examples. Madagascar's insect species are almost all island localized. Even the acridoids are 94% endemic (Preston-Mafham, 1991).

Recognition simply of areas with high percentage endemism need not be the only basis on which to determine conservation-worthy areas. However, K.S. Brown (1991) has emphasized that endemic epicentres can be monotonous, with low-level populations of genetically homogeneous insect species. Reserves should be large enough to cover both the nucleus and the peripheries of these endemic sinks, so that inner diversity and outer, dynamic diversity are supported. The inner areas are then essentially preserves for genetic uniqueness.

1.6 INSECT SIZE, PLAGUES AND POPULATION CRASHES

1.6.1 Insect size and conservation perceptions

Size is more than mere academic curiosity. It is the flamboyant flagships, particularly butterflies, that not only photograph well, catch the public eye and arouse sympathy, but are often the first species to appear on lists of protected insect species.

In size, insects are about midway within the animal kingdom. Man, in contrast, is a large animal, well within the top 1%. Insects to us are therefore small, which has important practical and ethical implications on perceptions for their conservation.

As only a small percentage of the tropical insect species have been described, any estimate of size pattern must involve substantial error (May, 1989; Stork, 1988). Specimens for museums are generally selectively collected along lines of human transportation and biased towards large size (Samways and Harz, 1982). Additionally, use of a single collecting method such as sweeping for comparative purposes (Schoener and Janzen, 1968) inherently misses the small, soil-surface species and may never catch, to name one group, the large evasive acridoids.

Although some of the largest and most frequently mounted species in public displays come from the tropics, May (1978) concluded that there is insufficient evidence to substantiate that there is any within-taxon general tendency for the larger size in the tropics.

Size in some cases relates to certain aspects of climate. Schoener and Janzen (1968) have suggested that larger size is associated with drier areas or those with longer growing seasons. Size also relates to resources. Davidson (1977) found that, in sites rich in seed-eating ant species, assemblages contain more very large species, which consume large seeds, and more very small species, which eat the small seeds. Interestingly, for some insect groups, e.g. female flower thrips, certain pollen beetles, tephritid flies, bird lice and fleas, there

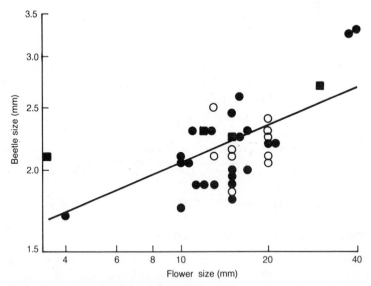

Figure 1.9 Body lengths of host-specific pollen beetles (*Meligethes*) plotted against host flower size, $\log_{10} Y = 0.192 \log_{10} X + 0.121$. The different symbols represent the different species groups. (From Kirk, 1991.)

is an increase in size with that of their flower or animal hosts (Kirk, 1991) (Figure1.9). The reason for this is not entirely clear, but may be due to various factors associated with the interstices of the host structure such as enemy-free space, access to more food, better microclimatic protection, etc. This further emphasizes the close relationship between insect variety and the character of the host habitat structure. The greatest number of Bornean canopy-inhabiting beetles are middle-sized and rare (Morse, Stork and Lawton, 1988) (Figure 1.10), and inhabit all parts of the plant structure.

1.6.2 From pest status to extinction

Although most of the few very large insects are associated with equatorial conditions, some Orthoptera in arid areas can be large. Some of these are locusts, being highly mobile and demographically highly variable, exploiting patchy resources which are often destroyed before the insects move on. These gregarious forms are not normally thought of as being of conservation status. Yet *Locusta migratoria*, a notorious gregarious pest in Africa and the Near East, is considered a threatened species in its solitary form at the edge of its range in some European countries.

Some of these orthopterans can have enormous population booms and crashes, from pest to extinction. The large wart-biter tettigoniid, *Decticus verrucivorus* ssp. *monspeliensis*, occurred in plagues in previous years this

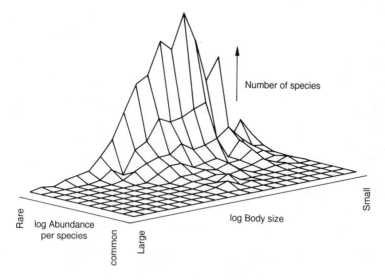

Figure 1.10 A three-dimensional graph plotting of arboreal Bornean beetle species against each log body length class and against total log abundance class for all beetles, and for each guild. (From Morse, Stork and Lawton, 1988.)

century and earlier, in the vicinity of Montpellier, southern France. This subspecies now may be extinct, last seen in 1972, despite intensive searches (Samways, 1989a). The reasons for the disappearance seem to be destruction of biotope to vine production, fragmentation of the landscape inhibiting swarming and possibly insecticide and fungicide usage. Selection pressure under this new landscape patterning seems to have favoured the solitary over the gregarious form. Incidentally, the solitary nominate species, *D. verrucivorus verrucivorus,* is a protected species in Britain and is sensitive to habitat structure modification (Shirt, 1987; Cherrill and Brown, 1990a,b). Yet it is common in many upland areas of Europe and across the Palaearctic region, where ancient meadows support it (Samways and Harz, 1982).

Insect conservation practices often emphasize large, rare population-stable, *K*-selected species. Yet the 'passenger pigeon phenomenon', whereby populations of an abundant species can collapse to nothing (albeit through hunting in the case of the pigeon), can be quite significant among insects. In the year 1478, locusts invaded the Venetian territory and produced famine. Such would be an extraordinary phenomenon today. Similarly, the Rocky Mountain grasshopper, *Melanoplus spretus,* which was a pest in the nineteenth century in the western states of North America, was extinct by 1902 (Lockwood and De Brey, 1990).

In terms of conservation biology, insects already have clearly indicated that great abundance at one time does not preclude the possibility of extinction in the future.

1.7 SUMMARY

Insects are numerous as individuals and as species, being by far the most dominant metazoan animal biomass, genetic variety and biotic species interactors in terrestrial ecosystems. Their structure, physiology and behaviour have been continually naturally selected to spawn a vast array of morphologies and life-history styles. Millions of niches have been cornered.

Insect structure is particularly suited to great variation, but restricts the animals to a small size. From a functional ecological viewpoint, there are many more insects than simply the number of species. Polymorphisms are common, with occasional great variation between the sexes, and often great differences between the young and adult stages. This developmental polymorphism may include similar modes of feeding but different media for the two stages, as in dragonflies. In the endopterygotes, the medium may be the same but the mode of feeding quite different, as in the butterfly. In insect conservation biology this variation in form and ecology must be taken into account.

The insect–plant interaction is the across-taxon most varied biotic interaction on earth. Insects are major players in ecosystem processes and plant survival. Insects are pollinators and nutrient cyclers, with the vast number of herbivores supporting many parasitoids, insect predators and mutualists, all of which require conservation.

Insect abundance is limited by abiotic and biotic factors. Mortality percentage in insects is high and variable, although offset by high natality. The complexity and variability, in time and space, of these mortality factors make predictions on the survival of even one well-researched species uncertain. This difficulty of prediction is particularly acute with the increasing anthropogenic disturbance and fragmentation of the landscape, whereby interspecific interaction strengths and types are altered with modified physical conditions.

Estimates of the number of described insect species vary from about 750 000 to about 1 million. One of the uncertainties about accuracy of these estimates concerns recognition of which species is actually a different one from another, as there are so many sibling species of insects.

Estimates of the number of total existing insect species also vary widely. Present suggestions range from 1.84 million to 50 million, with somewhere just under 10 million being the favoured figure.

Climatic adversity increases towards the polar regions, tracked by a general decrease in insect species richness. This decrease appears to be driven more by climate than by primary production in plants. Against a background of millions of years of lesser climatic adversity in the tropics, many plant species and complex plant architectures have evolved. These factors, coupled with plant–insect coevolutionary changes, and the effect of allopatric processes, have contributed to the vast insect variety in the tropics.

In the apparent changelessness of the tropical forests, there are daily or seasonal variations in weather, which affects insect seasonality. Insect population variability is as evident in the tropics as in temperate zones, making monitoring of insect species for presence/absence or abundance important at all latitudes.

Insect size has an anthropocentric element in that threatened species listings and population studies have been based mostly on large insects. Many insects, small and large, can outbreak to huge population levels. Among the larger insects are included some of the Lepidoptera and Orthopteroidea, which can cause considerable economic impact. Yet some of these plague insects can crash to zero population levels, and even become extinct as species, within just a few decades. In their solitary phase these economically important insects may be listed as threatened species in certain countries, yet elsewhere, in their gregarious phase, they may be rampant pests.

Insect conservation biology focuses on a vast array of variables – evolutionary, geographical and behavioural. These all have bearing on perspectives and action. As many and as varied as possible biotopes and landscapes need to be set aside. These reserve sites should also be as large as possible to accommodate the vagaries of often widely fluctuating insect population levels. The variety of sites should include areas for post-glacial faunas, insects of typical as well as unusual ecosystems, endemic-species refugia or sinks and species-dynamo areas.

-2

Past and present events leading to insect conservation concern

Greater dooms win greater destinies.

Heraclitus

Fritjof Capra once pointed out that the Chinese
ideogram for crisis, Wei Ji, is composed of two
interlocked characters. One is danger and the other is
opportunity. Thus the beauty of a crisis is that it holds
within its image inherent change.

Yatri

There are seven ages, each of which is separated from
the previous one by a world catastrophe.

Visud-dhi-Magga

2.1 PREHISTORICAL INSECT DISTRIBUTION IN TEMPERATE LANDS

2.1.1 Distinctions between north temperate and south temperate lands

There has been a tendency to view 'the north' as economically rich and relatively species poor and 'the south' (i.e. the tropics) as economically less privileged but rich in biodiversity. But 'the south' is more than simply the tropics. There are many historical, geographical and biological differences between the north temperate and the south temperate areas. There are also differences between the subtropical and tropical zones. These differences have arisen through prehistorical climatic events, isolation of many islands, and the relative size of land mass to sea, being much greater and more connected in the north than in the south.

The northern regions of Europe, Asia and America (northern Nearctic and northern Palaearctic) were depleted of their biota during the last ice ages. We are presently in an interglacial period, with insect species having moved northwards with the retreat of the ice and invading the thawed areas within the last 10 000 years (Mikkola, 1991) (Figure 2.1). Little speciation occurred

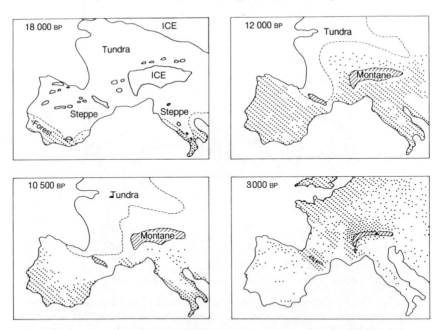

Figure 2.1 Proposed distribution of grasshoppers such as *Chorthippus parallelus* which survived in refugia in southern Spain and Italy at various stages in the post-glacial warming, including the 10 500 BP reversal. (From Hewitt, 1990.)

in the time available, although different subspecies have appeared. This has been visibly clear in morphologically distinctive and variable taxa. The British swallowtail butterfly (*Papilio machaon britannicus*) is easily separable from the continental subspecies. Other groups, such as the Odonata, have been morphologically more conservative, with British species generally identical to their continental conspecifics. Distinctive subspecies have had an important bearing on insect conservation, with range decrease or disappearance of endemic subspecies being of particular concern in Britain (Shirt, 1987).

2.1.2 Relicts

Immediately south of the glaciers, some species survived. They continue to do so today as glacial relicts in the cool mountainous terrains, as with the collembolan *Tetracanthella arctica* (Cassagnau, 1959) (Figure 2.2). These are interesting from a conservation viewpoint, as they represent examples of biodiversity created by glacial history (Mikkola, 1991). The threatened and

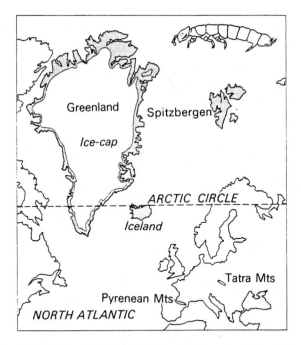

Figure 2.2 The distribution of the collembolan *Tetracanthella arctica*. It is found mostly in northern regions, but populations exist in the Pyrenees and in mountains in central Europe. These populations were isolated at these cold, high altitudes when the ice sheet retreated northwards at the end of the Pleistocene glaciation. (From Cox and Moore, 1985.)

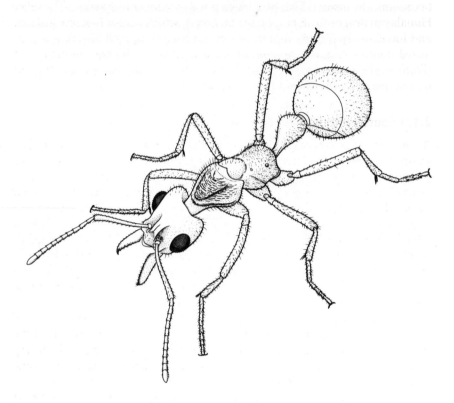

Figure 2.3 *Nothomyrmecia macrops*, an inhabitant of the Australian mallee ecosystem, is the most primitive living ant, known from elsewhere only from fossils 40 million years old. (From Hill and Michaelis, 1988. Drawn by S.P. Kim.)

now protected Apollo butterfly (*Parnassius apollo*) of Europe is a glacial relict showing different subspecies in the various mountain ranges. An action plan has been drawn up for its protection (New and Collins, 1991).

The geological long-term spatial dynamics is the backcloth for aspects of modern insect conservation biology. Some rare species today were formerly widespread in what are today quite disparate areas. Coope (1975) found that several coleopteran species presently living as narrow endemics in Siberia, Tibet and parts of Europe were common in the London area 100 000 years ago. It is uncertain whether they have shifted location or whether their ranges have contracted. The intervening environmental conditions have interplayed with the insects' dispersal, causing these range changes.

About 99.5% of all animals have become extinct naturally throughout life's history (Simpson, 1952). Some species of insects are the last relics of a once abundant taxon. Often these are geographically isolated, as well as being taxonomically insular. This provides a good case for conservation. The relict Himalayan dragonfly (*Epiophlebia laidlawi*), which occurs in a few isolated and increasingly disturbed biotopes in northern India and Nepal, is a protected species of great taxonomic curiosity. Similarly, the dinosaur ant (*Nothomyrmecia macrops*), which is the most primitive living ant, is of great taxonomic and behavioural value (Figure 2.3).

2.1.3 South temperate areas

At the time of northern glacial periods, the earth was, overall, cooler. The southern hemisphere experienced isolated periglacial conditions, but not to the extent of eliminating biota as in the far northern hemisphere. Temperatures in southern Africa dropped to 8°C below those of the present day. Biota survived and speciated with changing climatic conditions. Today this relict vegetation is seen in the 23 200 species of plants in southern Africa, 80% of which are endemic. The species richness of the southern Cape (calculated as species/area ratio) is the highest in the world, 1.7 times that of Brazil (Davis *et al.*, 1986). Many species have survived the last 9000 years of rapid changes in temperature and moisture in isolated areas of hillsides and in mountain crevices.

Across the Indian Ocean, with the exception of the southern part of New Zealand, conditions were moister 6000 years ago than today. Much of the Australian fauna, including the insects, has adapted to an increasingly dry and fire-prone landscape (New, 1984). These historical factors pose some interesting conservation questions for Australia's insect fauna. At one extreme there are biologically important individual species of highly restricted distribution, and at the other, there are diverse radiations often covering the whole continent in their range (Greenslade and New, 1991).

2.2 PREHISTORICAL SETTING IN TROPICAL LANDS

2.2.1 Past tropical diversity

The Pleistocene and Holocene climatic changes affected the geographical distribution of biotas, leaving refuge areas far less extensive than those of today (Connor, 1986). Detailed geographic evidence describing how the climate of the lowland tropics changed is generally lacking. But against a background of relative climatic constancy, albeit generally cooler than today, major patterns of speciation and biogeography may have been determined as much by forest fires and river dynamics (Sanford *et al.*, 1985; Salo *et al.*, 1986) as by Pleistocene climatic changes. The world biota was more compressed into the lower latitudes during the Pleistocene glacials than interglacials. There is virtually no evidence on how these freeze compressions

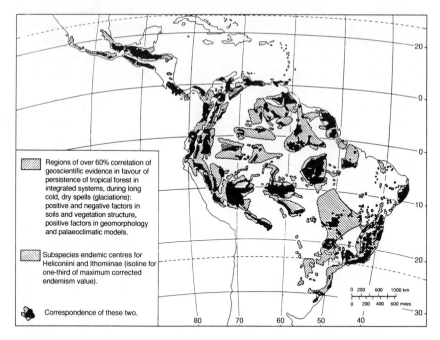

Figure 2.4 Geographical overlay of Heliconiini and Ithomiinae butterfly subspecies-endemic centres with geoscientific model of probability (more than 60% in synthesis of palaeoclimatic, geomorphological, paedological and phytosocial data) for forest permanence during the most recent long cold dry climate spell (Würm–Wisconsin glaciation, 12 000–20 000 years BP), based on observations of present-day forest in regions of such unfavourable climate in Brazil. Essentially all endemic centres enclose areas of high probability for existence of 'palaeoecological forest refuges'. (From K.S. Brown, 1991.)

influenced insect species richness, but as plant richness remained high in these areas presumably so did that of insects. This is supported by K.S. Brown's (1982, 1991) data on heliconiine and ithomiine butterflies in the neotropics (Figure 2.4).

Many plant–insect associations are non-interactive (Strong, Lawton and Southwood, 1984), and many insects attack more than one plant. Insects have diversified possibly more through isolation effects than through a coevolutionary relationship between plants and insects, although both these factors, and others, have all contributed to the great insect variety (section 1.5). The importance of insularity is evidenced by the patchy distribution of plant and insect species in what appears otherwise to be a continuous tropical forest. Erwin (1988) has pointed out that beetle samples collected in four different forests in the central Amazon showed that 83% of the species were

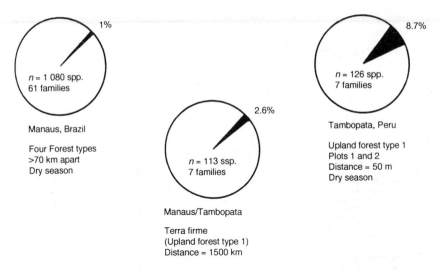

Figure 2.5 Pie diagrams of shared beetle species among forests in Peru and Brazil in percentage of fauna. (From Erwin, 1988.)

found only in the samples of one of the types of forest, 14% of the species were shared between two and only 1% were found in all four forest types. In Amazonian Peru, he compared two sites in the same forest type, separated by only 50 m, and found that only 8.7% of the beetle species were shared between the two sites (Figure 2.5).

2.2.2 Extrapolation from marine data

Global species richness, as measured by marine fauna, which has been fossilized and is readily visible, has continued to soar during the last 100 million years (Benton, 1986). A similar, but not such extreme trend is seen in genera and families, although orders appear to have remained fairly steady for the last 450 million years.

The present biodiversity crisis represents a sudden drop in biotic diversity, particularly at the subspecies and species taxonomic levels. Extinction spasms have occurred in the past. It is not certain exactly how these records, based principally on marine invertebrate calcareous skeletons and terrestrial reptile fossilized remains, really reflect the overall trend in earth's species richness patterns. Briggs (1991) suggests that general biodiversity was virtually unaffected, as most of this was, and is, composed of terrestrial arthropods and plants, which appear not to have undergone the same extinction spasms.

The feature of today's extinctions rate is that it is being accelerated by man's activities. It is possibly 1000 to 10 000 times faster than would have occurred naturally. Insects have a high profile in this conservation concern because they are the taxon under greatest threat, in terms of numbers of species likely to become extinct.

2.3 HISTORICAL TRENDS IN TEMPERATE LANDS

2.3.1 Early fragmentation of the landscape

Man's impact on the land surface has been principally over the last 10 000 years, particularly in the Middle East. One of the first urban settlements was Catal Hüyük, in Anatolia, thought to be built about 7200 BC. Such settle-

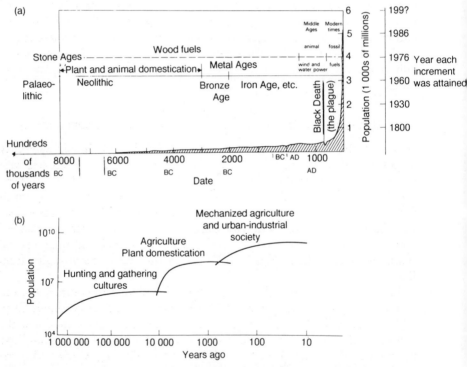

Figure 2.6 (a) At the time of the appearance of settled agriculture the human population growth rate was relatively low but nevertheless increasing. It took hundreds of thousands of years for the numbers of humans to reach the first 1000 million. But the second, third, fourth, and fifth increments of 1000 million took 130, 30, 16, and 10 years respectively. (b) This logarithmic plot assumes that technical change enhances human population growth but that it then adjusts to a new equilibrium, only to be disrupted again by changing technology. (From Mielke, 1989.)

ments were the start of human fragmentation of the landscape, which has continued at an increasingly fast pace (Figure 2.6). Catal Hüyük is on the Konya Plain, which had been a lake up until a few thousand years prior to the town. This left fertile soil and the availability of water. In the surrounding Taurus Mountains were game and timber resources. This situation is an example of nodular fragmentation (i.e. increasing degradation at the centre of a site). In its early stages, it must have encouraged edge species, with perhaps even some of the first pest species appearing. Certainly there are millennia-old records of locust invasions and insect pests of stored grain.

Since these earliest settlements, the impact of man on natural ecosystems in temperate lands has varied according to geographical location. Types of agricultural development, perceptions on harmony with the earth and human migrations have all played significant roles. These in turn have been swayed by growth in human populations and clashes of cultures. This has been particularly evident in the Americas, where the exploitative approach of mechanized agriculture has clashed with the ecologically sound, environmentally harmonized approach to land that was part of the North American Indian way of life.

2.3.2 Differences between the northern and southern hemispheres

Until recently, much of far northern Europe, Russia and Canada had fairly large tracts of undisturbed land, providing refuge for many vertebrates with large home ranges. In the southern hemisphere, the pampas of the southern tip of South America has been only recently disturbed. Southern Africa, which for many centuries supported only the Khoi-Khoin (Hottentot) pastoralists and the San (Bushman) hunter–gatherers, was also relatively undisturbed until the arrival of the negroid settlers from the north, and then, several centuries later, by the Caucasian settlers from the south after voyaging from Europe.

In New Zealand, the earliest migrants arrived in the tenth century AD, and extinctions of larger animals soon followed, as on many oceanic islands (Day, 1989). The Australian aborigines seem to have extensively modified the landscape using fire for many centuries. Such man-initiated fires are additive upon naturally caused fires, especially through lightning, which strikes the earth 100 000 times each day (Goudie, 1989).

2.3.3 Landscape fragmentation and woodlands loss

The Mediterranean (Naveh and Lieberman, 1990), the Middle East and, later, northern Europe have undergone dramatic landscape changes throughout historical time. Neolithic man undertook the first forest clearances. At that time (3000 BC), Britain was covered by a rich and extensive variety of trees, known as wildwoods (Rackham, 1986). Earliest settled agriculture of Europe involved localized cropping between these wildwoods until relative soil exhaustion. Such activities would have benefited edge species, and may well have provided greater opportunities for larger populations of flowering herbaceous plants. These plants, in turn, provided greater nectar sources, which, for butterflies at least, are often a limiting factor (Warren, 1985).

In these agricultural patches, there was little rotation of crops, no alternating tillage with properly prepared soil, nor any naked fallow system. Presumably on abandonment, these patches reverted to succession starting with grassland. This almost certainly could have benefited certain species in such groups as the acridoids, lycaenids and cercopids, especially where the land was grazed, which probably encouraged a variety of herbaceous plants.

By AD 1086, the matrix of trees had become agricultural land, with the English lowlands comprising about 7% woodland, made up of wildwood, woodpasture and coppice. Again this must have changed the overall species balance of insect communities enormously. Today only remnant wildwood patches remain, with 50% of that remaining lost between 1947 and 1980 alone. Tree removal continues, with only 0.48% of the land surface of Lincolnshire now being covered with wildwood. Opening of the landscape not only increases the edge effect in terms of succession, but also exposes the opened patches to increased battering from heavy storms, such as those of the late 1980s and 1990. In addition, there has been increasingly extensive planting of high coniferous forest, with a decrease in managed coppice-wood, both of which have been detrimental to local insect species richness in Britain (Figure 2.7).

The Anglo-Saxons took over England as a partly hedged landscape (Rackham, 1980). Medieval England had essentially two types of landscape, the ancient countryside and the planned countryside, which may have their roots as long ago as the late Roman period (Rackham, 1986). The ancient countryside is a general piecemeal growth and change over the centuries, with hamlets, lonely farmsteads, winding lanes, dark hollow-ways, intricate footpaths, thick mixed hedges and many small woods. Such a countryside reappears in Europe, in the *bocage* of Normandy, and in parts of Greece and Italy (Zanaboni and Lorenzoni, 1989). In contrast, the planned countryside was of large regular fields with flimsy hawthorn hedges, of few, often straight, roads, of clumps of trees in the corner of fields and of a large village every 2 miles or so. Although we have no records, such landscape changes must have greatly influenced the local distribution of insect species.

These landscape changes probably interacted with increased temperatures, which in Britain were 2–3°C warmer 10 000 BP to 4500 BP (Dennis, 1977). The earliest wildwoods probably had many temporary gaps, possibly at least partly mammal induced. These temperature and gap factors would have

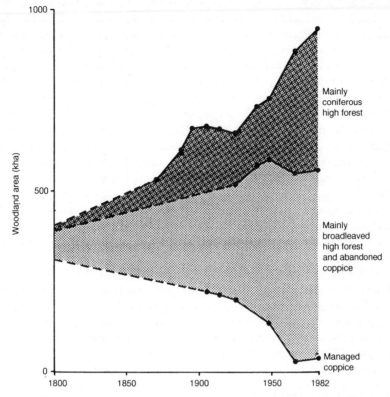

Figure 2.7 Changes in the extent and composition of woodland in England since the year 1800. (From Warren and Key, 1991.)

influenced the relative abundances and distributions of British butterflies with different biotope preferences. Man's clearances and land management may have encouraged those species restricted to warmer microclimates of disturbed woodlands, grasslands and heaths. This would have continued until the present century. The new woodland management practices, through more closed canopies, caused the disappearance of warm microclimates (prior to global warming) from most seminatural biotopes (J.A. Thomas, 1991).

2.3.4 Hedges

Fields, meadows and hedges for many centuries have continued to replace woodlands. These hedges came to support a characteristic fauna depending on geographical location, age and structure (Pollard, Hooper and Moore, 1974). In recent years, and with increasing emphasis on farm efficiency and government incentives to increase agricultural produce, hedges have been removed increasingly intensively in the second half of this century. By the mid-1960s, 8000 km of hedgerows were being removed annually. In Huntingdonshire, England, by 1972 only 20% of hedgerow trees present in 1945 still remained (Pye-Smith and Rose, 1984). This trend has slowed or halted with the introduction of incentives geared more to conservation of wildlife than in the past.

2.3.5 Stone walls

In the appropriate geological areas of Europe, some stone walls have been in existence for many centuries, and today have a rich and varied biota (Darlington, 1981). Unlike hedges, there has been negligible wall removal. Walls act as stone dumps as well as barriers, and there is little point in redispersal of the construction materials. Also, many of the extensive ancient walls are in grazing rather than crop areas, and there has not been the same economic incentive to remove them.

2.3.6 Meadows

Meadows, which were an integral part of the ancient and medieval village, are rich and varied in biota (Morris, 1971). Meadow loss has been substantial this century. In Finland, the proportion of meadows was 62% of the permanent agricultural land in 1880, but then declined to 29% in 1920, and is negligible today (Hanski and Tiainen, 1988).

2.3.7 Wetlands and heaths

The loss of wetlands and heaths has also continued intensively and rapidly. Indeed, the loss of wetlands has been not only a temperate problem but also a tropical one (Sioli,1986). In Lancashire alone, 99.5% of the lowland bogs have been lost, while the fens of East Anglia are reduced to mere remnants, such as Woodwalton Fen and Wicken Fen, which have to be intensively managed, particularly hydrologically. Similarly, the loss of heathland has been severe. In Dorset, England, 75% of the heathland has gone in the last 50 years, and 76% of the Scottish Fife heaths have disappeared since the 1950s. Of the upland heaths, 90% were lost from Dumfries and Galloway in the same period, while farther south the lowland Brecks of Suffolk, England, 70% has been converted to mostly cereal fields (Pye-Smith and Rose, 1984).

2.4 HISTORICAL TRENDS IN TROPICAL LANDS

Throughout man's recorded history, shortage of resources has caused population declines and migrations (Whittaker, 1975). Although Pleistocene man was probably widespread, his impact on the land surface was small until the formation of villages in many parts of the world about 7000–6000 years BC. The rapid human population growth in the tropics today is principally from either recent immigrant descendants, as in South America, or more established cultures, as in Indo-China. Cultural perceptions on family size coupled with improvements in hygiene and insect vector control (Pimentel, 1975), have also contributed to the increase.

Whereas for centuries the equatorial forest had supported a low population of semi-itinerant hunter–gatherers and shifting cultivators, the increased population level has demanded more intensive and expansive shifting agriculture, coupled in recent years with increasingly extensive clear felling for rapid economic return (Myers, 1979; Collins, 1990). In Rondonia, one of the southern Amazonian states of Brazil, there has been a 10-fold increase in human population between 1975 and 1986, from 110 000 people to over 1 million. In 1975, almost 1250 km of forest were cleared. By 1982 this figure had reached 10 000 km^2, and in 1985 it was around 17 000 km^2 (Fearnside, 1986). For insects, this massive landscape modification has been devastating in fragmenting populations and causing species extinctions. But the actual number of species extinctions is far from being clear. The main point is that in the tropics the level of species endemism, particularly on islands, is very high compared with the north temperate lands, and the impacts have been much more severe and rapid.

Regarding islands, Howarth and Ramsay (1991) have pointed out that 90% of terrestrial arthropods are endemic to New Zealand, whereas for Hawaii this figure is as high as 99%, and man's impact has been devastating as a result of landscape change through agriculture and urbanization, and the accidental and deliberate introduction of invasive biota.

2.5 RECENT GLOBAL CLIMATE CHANGES

2.5.1 Enhanced global warming

The earth has been warming for the last 15 000 years (Woodwell and Ramakrishna, 1989). In recent years this warming has accelerated due to the continuing accumulation of heat-trapping gases, such as carbon dioxide, methane, nitrous oxide and chlorofluorocarbons (CFCs) (Lyman, 1990). Over the last century, the amount of carbon dioxide in the atmosphere has increased by more than 25% as a result of deforestation and combustion of fossil levels. The rapidity of the recent changes has not been conclusively proved, but may be extremely rapid, and will probably remain unproven until after the effects have been felt (Woodwell and Ramakrishna, 1989). This artificial warming rate could be 100 times faster than any natural increase. Also, there is no sign of the warming trend letting up, despite cooling natural factors, such as low levels of solar radiation, high levels of volcanic activity, shading and increased dust and sulphate particles in the air. A warming of 2°C will make the earth warmer than at any other time in written human history, and an increase of 4–5°C will make it warmer than any time in the last one million years. Estimates vary as to the extent of the present warming trend, with an overall global average possibly of 0.8°C per decade. High-latitude areas would experience rises higher than the global average, with polar regions warming by two to three times the global average. Warming in the tropics would be limited to 50–75% of the average (Lyman, 1990).

Most workers expect a global temperature increase of a few degrees centigrade over the next 50–100 years which will cause the sea level to rise by between 20 cm and 1.5 m as a result of thermal expansion of the oceans, the melting of montane glaciers and the possible retreat of the Greenland ice sheet's southern margins (Schneider, 1989). In Antarctica, ice may actually build up, owing to warmer winters, which would probably increase snowfall.

Such warming effects will speed biotic metabolism, especially respiration, with an increase of 1°C increasing the rate of respiration of plants and the decay of organic matter in soils by 10–30% or more (Woodwell and Ramakrishna, 1989). Being ectotherms, the same effects will impact on insect metabolism directly, as well as having indirect effects on their biotopes.

An additional factor is that temperature falls by 0.6°C for every 100 m rise in elevation, which suggests that to maintain the same local thermal environment, and taking the possible overall global average, species theoretically would have to move up the mountainside by at least 500 m by the middle of the next century. One problem is that as mountains are roughly conical, the land areas become less with elevational increase. Yet to stay at the same altitude, using the broad rules of ecosystem change with latitude (MacArthur, 1972), species would have to move by about 250 km towards the poles. Insect movement and establishment would be inhibited less by the rapidity of global warming than by the increased fragmentation and modification of the landscape. Such fragmentation would prevent the normal smooth population drift.

2.5.2 Other global influences

Besides a general enhanced global warming, it is projected that climates will be more variable, with a possible increase in inclement conditions, such as droughts, hailstorms and hurricanes. Some subtropical areas such as South Africa, with an increasingly drier climate, will suffer soil moisture conditions 11–18% lower than at present (Huntley, Seigfried and Sunter, 1989). The many range-restricted endemic insect species would suffer greatly.

Other changes in contemporary climate include enhanced acid deposition by rain, urban fog and thinning of the stratospheric ozone (O_3) shield that protects the earth from harmful ultraviolet radiation (Graedel and Crutzen, 1989). Hyperacid deposition is already well known to have affected many parts of the northern hemisphere in particular, causing tree dieback, acidification of freshwater rivers and lakes and changes in the paedosphere. Some insect species may benefit from changes wrought by acid precipitation, but, in general, insect diversity will almost certainly decline. This will be caused either by a direct effect on the insects themselves or by loss of the plants on which they depend. These sorts of impacts are urgently in need of further research.

In the earlier stages of particulate airborne industrial pollution there was an increase in melanism among certain insects. The best-studied case is the peppered moth (*Biston betularia*), found in Britain, whose proportion of melanic to non-melanic forms increased with the increase in the deposition of dark-coloured industrial deposits on trees. With the reduction in air pollution of a deposition nature, there has also been a decrease in the level of melanism in these moths, but the situation is complex and closely associated with the resting behaviour of the adults (Brakefield, 1987).

The Chernobyl nuclear reactor accident, in the Ukraine in 1986, led to migrant moths carrying radioactive particles as far as Finland, while some local Finnish moths picked up radiation from substrate surfaces (Mikkola and Albrecht, 1986). Indirect effects of the global changes on insects through impact on plants and ecosystems are not clear at this stage, but simulations on plants show that elevated carbon dioxide levels on transpiration and gas exchange will increase the sensitivity of community structure, particularly that of forests (Woodward, 1990).

Parsons (1990) points out that the multiple effect of the global stressful conditions may, at the limits of species' geographical ranges, impose physiological constraints, with the amount of energy required and stress being too great to permit survival. This stress would be both on the herbivorous insects as well as on their food plants, with a possible cascade effect on ecological dependants such as parasitoids, hyperparasitoids and mutualists.

There are suggestions that climatic change will not be all bad for certain insect taxa. Also, there will be indirect effects through impacts on the host plant as well as direct impacts on the various life stages of the insect. In a predictive study on the effect that climatic change may have on British butterflies, Dennis and Shreeve (1991) suggest that the adverse impact will be particularly strong when mediated through the larval host plant. The impact will vary from one region to another. Climatic warming may reduce the regional contrasts in species diversity and population status between northern and southern Britain. There may be range expansions and upland colonization in the north, while there may be reduced populations and some population extinction in the south. Climatic warming also may encourage new residents and migrants in southern Britain from continental Europe.

Often it is difficult to prove retrospectively that localized pollution has caused a decline in insect numbers or species. Toxic effluent is recognizable by the differences in insect communities above and below the industrial plant. In Britain, the extinction of the orange-spotted dragonfly (*Oxygastra curtisii*) co-incided with the opening of a new sewage treatment plant just upstream of its last known habitat (Fry and Lonsdale, 1991). The combinations of stress conditions vary around the globe, and occur in varying and overlapping scales. Severe local acid precipitation combined with, say, global warming may be synergistic in central Europe. At the tips of southern continents, thinning of the ozone layer, combined with global warming, may be the biggest threat. Little is known at present of the impact that the ozone thinning may have on plant and insect communities, particularly as it is a variable phenomenon from year to year. In 1988 for instance the ozone hole represented a depletion of about 30%, while in 1989 it was up to 45% (Scourfield *et al.*, 1990). Some eurotopic species will expand their ranges. In other species populations will shrink and fragment into meta-populations. For stenotopic types, the species may not survive at all.

2.5.3 Interplay between historical, local and global impacts

Historical biogeographical factors affect each other and influence present conservation decisions. Prehistorical events, such as glaciations, have given the north-temperate zone quite different species endemism and distributional profiles from those of the south-temperate areas. Chown (1990) has illustrated how more ancient events, such as quaternary climatic events, have influenced the insect communities on the sub-Antarctic South Indian Ocean Province Islands. The past history of each particular island has determined its present insect species composition, with both higher plants and insects surviving on the islands during the glacial maxima. At equivalent latitudes in the northern hemisphere (i.e. about 50° from the equator) the glacial fronts removed the biota.

From a biogeographical perspective, besides prehistorical events, there also has been larger and extensive human landscape modification in the north-temperate zone. In historical times, agricultural and industrial events have had an increasingly large-scale effect. In earlier historical times the landscape was modified by low-energy agriculture. The scale of modification is now global, not only through more intensive and extensive modifications of the local landscape, whether through removal of hedgerows in Europe or wholesale removal of tropical forests, but also because of enhanced climatic changes. All add up to landscape modification and destruction to the detriment of communities of which insects are the major faunistic component. The twentieth-century global impacts have varying effects across the world, and global management requires not only international protocols reducing global impacts, but localized land management in response to them and to other, more localized, impacts on the landscape.

2.6 SUMMARY

A knowledge of the past has great bearing on management for the present and future. Both prehistorical and historical events have had considerable influence on insect species' large-scale and small-scale distributions and abundances. The Pleistocene glaciations were particularly significant in causing depletion of faunas in the northern hemisphere, leaving today areas such as Britain and Fenno-Scandinavia with very few insect endemics. Although the influence of the glaciations caused a worldwide decrease in temperature, the cooling allowed many taxa at the equivalent latitudes in the southern hemisphere to survive and speciate, making the temperate southern hemisphere genetically rich floristically and entomologically. The tropical areas receded to 'Pleistocene refugia', with the forests expanding again after warming had begun.

In historical times, man's impact upon the Palaearctic north-temperate landscape has been accelerating and increasingly fragmentary, to the detriment of interior woodland species. Nevertheless, the management practices of grasslands, woodlands and hedges has given the opportunity for many species to survive and thrive. With the onset of intensively mechanized agriculture these species have declined along with their habitats.

Intensive habitat loss is now worldwide and there is particular concern regarding loss of the species-rich tropical biotopes. Concern also concentrates on the tropics, as their destruction is adding to global warming. The effect will be particularly severe on tropical insects that are not able to tolerate great variations in climate.

Global warming is compounded by other environmental pollution effects, including enhanced ultraviolet radiation through thinning of the stratospheric ozone shield and hyperacid precipitation. The impact on plant life is already locally severe, and is likely to have an increasing effect on insect life. Such effects are inescapable. Global warming and the increasingly fragmented and degraded landscape, the rapidity and magnitude of which have become so detrimental, are not likely to allow insects and their plant hosts the opportunity to move up in elevation or across latitudes, as they may have done during past, more gradual, natural climatic changes. Global management involves not only large-scale international protocols to reduce adverse global impacts, but also localized land management plans to deal with the interacting local impacts.

3

Emergence of insect conservation biology

AD 1872 Then appear the Bath Whites, Queens of
Spain, and Camberwell Beauties, the last unusually
plentiful and extending to Scotland.

A.H. Swinton (c. 1880)

Before I flew I was already aware of how small and
vulnerable our planet is: but only when I saw it from
space, in all its ineffable beauty and fragility, did I
realize that humankind's most urgent task is to cherish
and preserve it for future generations.

Astronaut Sigmund Jöhn

3.1 DEVELOPMENT OF INSECT CONSERVATION CONCERN

3.1.1 Early history

Pyle, Bentzien and Opler (1981) have given a detailed history of modern practical and legislative insect conservation which began in the first half of the last century, when in 1835 the Apollo butterfly *(Parnassius apollo)* was given protection under Bavarian State Decree. Concern was expressed in Britain that the large copper butterfly (*Lycaena dispar dispar*) was extinct by the middle of the nineteenth century. The decline of other British butterflies stimulated further awareness, with Charles Rothschild recognizing the importance of protecting their habitats and special biotopes, rather than individual rare species threatened with extinction. The tide of concern led to the founding of the Society for the Promotion of Nature Reserves in 1912. Over the next few years, 284 sites were identified, including many that contained rare insects.

3.1.2 Influential modern societies

The Insect Protection Committee of the Royal Entomological Society came into being in 1925, and developed into the Joint Committee for the Conservation of British Insects. This committee is highly active today (e.g. Collins, 1991), alongside other groups such as the Invertebrate Working Group of the National Zoo Federation. In France, l'Office pour l'Information Eco-entomologique is also active, producing regular communications. In the USA, The Xerces Society was founded by Robert Pyle in 1971, and is named after the Californian butterfly, *Glaucopsyche xerces*, last seen in 1944. This society produces an inspiring invertebrate conservation essay journal, *Wings*, and has now become truly international, with wide readership and active research projects, for example in Madagascar, a centre of megadiversity, and where 77% of the butterflies are endemic.

3.1.3 Major international organizations

Of particular significance are the increasing invertebrate activities of the IUCN (the World Conservation Union, formerly the International Union of Conservation of Nature and Natural Resources). The IUCN has a highly influential membership, including 60 governments and more than 500 non-governmental organizations among its 650 members from 120 countries. The Species Survival Commission of the IUCN has several Specialist Groups: social insects, Lepidoptera, Odonata and, most recently, Orthopteroidea and Water Beetles. Other groups are being planned

Major conservation projects are undertaken by the World Wide Fund for

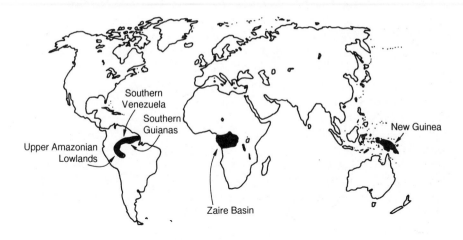

Figure 3.1 The remaining wilderness areas where there are large tracts of primary rain forest rich in insect species and other biota (From McNeely *et al.*, 1990. After Conservation International, 1990.)

nature (WWF), The World Bank, the World Resources Institute and Conservation International (see Durrell, 1986; Reid and Miller, 1989; McNeely *et al.*, 1990), as well as other organizations. The thrust is on biodiversity conservation, which is the conservation of genes, species, ecosystems and landscapes. Insects, being so numerous and ecologically important, fall very much under the ambit of these wider conservation contexts. Emphasis is on policies and projects that conserve as much of the biota as possible. These policies include identifying and designing conservation plans for major tropical wilderness areas (the few remaining parts of the world where large tracts of primary tropical forest still exist, e.g. the southern Guianas–southern Venezuela–Upper Amazon Lowlands belt, Zaire Basin and New Guinea) (Figure 3.1) and critical or 'hotspot' areas where there is high biodiversity but considerable human population pressure (e.g. Colombian Chocó–western Ecuador–uplands of Western Amazonia belt, eastern Madagascar, eastern Himalayas, peninsular Malaysia, northern Borneo, Philippines, Queensland, New Caledonia and Hawaii) (Figure 3.2) (Myers, 1988, 1990). From a more politically boundary-defined viewpoint, it appears that 50–80% of the world's biological diversity will be found in six to twelve tropical countries, the first six of which are Brazil, Colombia, Mexico, Zaire, Madagascar and Indonesia (Mittermeier, 1988) (Figure 3.3).

Returning specifically to insects, in Europe in particular there are many other interested groups and societies devoted wholly or partly to insect conservation. One is the British Butterfly Conservation Society formed in 1978.

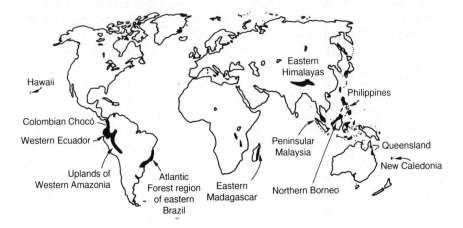

Figure 3.2 Critical areas of the world, or 'hotspots', which are rich in insect and other biotic diversity yet are under pressure from the human population. (From McNeely *et al.*, 1990. After Myers, 1988.)

The above organizations are devoted to protection in the field, of both species and biotopes. The Convention on International Trade in Endangered Species of Wild Fauna and Flora (CITES) legislates trading restriction of threatened species. In the 1989 update of the Appendices to the Convention, the butterflies *Ornithoptera alexandrae, Papilio chikae, P. homerus* and *P. hospiton* are listed under Appendix I, that is species threatened with extinc-

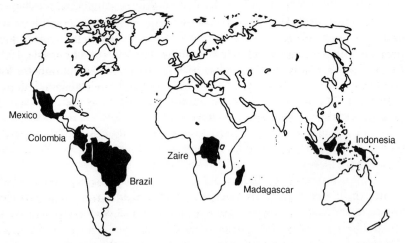

Figure 3.3 The six richest countries in biodiversity in the world as identified by the World Wide Fund for Nature. (After Mittermeier, 1988.)

tion through trade. Appendix II, which lists species that may become threatened with extinction if trade involving them is not regulated, includes the butterflies *Bhutanitis* spp., *Ornithoptera* spp., *Parnassius apollo, Teinopalpus* spp., *Trogonoptera* spp. and *Troides* spp. These lists are likely to increase in the future.

CITES is not involved in active on-site protection of populations of species, which is controlled by local and national conservation authorities. Ironically, by being CITES listed, species increase in their rarity and commercial value, which in the case of some vertebrates has increased demand for them from collectors and traders (Joyce, 1989). This is also true for the invertebrates, and sometimes listing them as rarities has stimulated increased activity by unscrupulous collectors. Confidentiality of sites, protection in reserves and goodwill among members of entomological societies go a long way to circumventing this problem.

3.1.4 National activities

Many countries have national and state or provincial regulations to protect listed insects or their habitats. The 1973 United States Endangered Species Act has led to some detailed recovery plans (Opler, 1991). However, there has been some recent criticism of the Act because it fails to account accurately for important biological concepts such as ecosystem conservation, patch dynamics and the probabilistic nature of stochastic threats to a species' persistence (Rohlf, 1991). Bearing in mind that insects generally have small home ranges, yet also cautioning against inclement weather conditions, many city and town parks (Gilbert, 1989) and botanic gardens provide refuge for insects (Davis, 1978; Samways, 1989b). Ecological design must integrate with aesthetics and protection for the area to be a significant, albeit local, protective area for insects (Goode and Smart, 1986; Fry and Lonsdale, 1991). In the short term it is vital that these reserves are managed not only for local conditions, but also bearing in mind future global environmental changes (Morris, 1991). But reserves, however well managed, do not guarantee the survival of insect species (J.A. Thomas, 1991).

Inadvertent insect conservation where sites are military, archaeological or water catchment areas is also playing a major role in many countries (Samways, 1989c).

Setting aside of specific reserves for a specific insect species is almost non-existent. One rare example is the 12-ha Ruimsig Entomological Reserve, South Africa, which was established among housing development by the Roodepoort City Council specifically for the threatened lycaenid butterfly *Aloeides dentatis dentatis* (Henning and Henning, 1985).

3.2 PERSPECTIVES ON INSECT CONSERVATION

3.2.1 North-temperate perspectives

Insect conservation approaches over the last 100 years or so have changed. In the last century, the main theme was disappearing British Lepidoptera. Earlier this century came the realization that insects are important components in ecosystem function, from nutrient turnover and energy transfer to pollination. Although the species richness of the tropics was recognized over 150 years ago, when Duncan (1840) provided tables of the comparative richness of Coleoptera from the north-temperate regions to the tropics, it is the tropical forest destruction that has highlighted the plight of so many insect species and the places where they live.

There has been an increasing awareness of the preservation (remaining intact with no management) and conservation (maintaining intact with management) of the insects' biotopes as a means of individual species preservation (Pyle, Bentzien and Opler, 1981; New, 1984; Collins and Morris, 1985; Collins and Thomas, 1991; Fry and Lonsdale, 1991; Spellerberg, 1991).

The importance of maintaining the last remaining remnants of natural ecosystems in human population-dense Europe (e.g. Moore, 1987a; van Tol and Verdonk, 1988) and North America (Opler, 1981) has led to the setting aside of some specific reserves, with insects being among the primary, if not main, subjects. The Suffolk Wildlife Trust, England, has purchased a 100-acre Site of Special Scientific Interest near Oulton Broad, with the aim of saving the Norfolk hawker dragonfly (*Aeshna isosceles*).

The fragmentation of the north-temperate landscape has triggered an awareness of this patterning in conservation approaches. Elton (1966) recognized this when he described the interstices between agricultural land as being important reservoirs for biota. The scientific field of landscape modification, patterning and function has now been formalized into 'landscape ecology' (Forman and Godron, 1986). This is no longer simply the domain of the geographer, but also that of the insect conservationist (Samways, 1989c; Bunce and Howard, 1990; Fry, 1991).

3.2.2 Tropical and south-temperate perspectives

The task of insect conservation in the tropics is of quite a different dimension altogether from that in the north-temperate areas. With possibly less than 5% of the vast insect numbers scientifically named, insect conservation in these tropical zones falls under the overall umbrella of biodiversity conservation. Costa Rica, to name one example, has been highly active in preserving representative ecosystems (Janzen, 1989), although the natural history devastation has nevertheless already been great, with 70% of its indigenous ecosystems having been greatly disturbed (Ugalde, 1989).

There are some instances in the tropics where highly prized, glamorous and geographically restricted species have specifically triggered attempts at specific habitat conservation. The magnificent Queen Alexandra's birdwing butterfly (*Ornithoptera alexandrae*) inhabits primary and advanced second-

Figure 3.4 Distribution of Queen Alexandra's birdwing butterfly (*Ornithoptera alexandrae*) in Papua New Guinea. Highland areas over 1000 m above sea level (a.s.l.) are stippled. (From Parsons, 1984.) The butterfly is classed as 'Endangered' by the World Conservation Union (From Wells, Pyle and Collins, 1983.)

ary lowland rain forest in the small Popondetta Plain area in the Northern Province of Papua New Guinea. Attempts are being made to preserve its habitat (M.J. Parsons, 1984; Collins and Morris, 1985) (Figure 3.4).

With the inauguration of the World Conservation Union's (IUCN), Species Survival Commission and the formation of several specialist groups, individual tropical species conservation is also being recognized. The Lepidoptera Specialist group has produced the first Action Plan – in this case for swallowtail butterflies (Papilionidae)(New and Collins, 1991). Even with this high-profile taxon, it is evident that there is still considerable knowledge lacking on distribution and basic biology of many important species. Such Action Plans are useful not only in pinpointing which species require research, but also in stimulating research on how landscape conservation, both in and out of nature reserves, is influencing the species' demographic changes.

3.2.3 The initial alert to insect species loss

Loss of the British large copper butterfly (*Lycaena dispar dispar*), last seen in 1865 (Shirt, 1987), was through draining and cultivation of its fenland breeding sites, such as Whittlesey and Yaxley Meres (South, 1906). Loss of the xerces blue butterfly (*Glaucopsyche xerces*) was also due to habitat loss, but in this case under the expanding outskirts of the city of San Francisco (Murphy, 1988). In both cases, concern was raised, and rightly so, that such highly visible and aesthetic insects in economically rich socioeconomic settings should disappear. The reality today is no different, only the magnitude of species loss through landscape modification is far greater, especially in the underprivileged socioeconomic conditions of the tropical countries. World over, it is a salvage operation, but it is the genetic loss in the low latitudes that is particularly acute.

3.3 INSECT CONSERVATION AND PLANET MANAGEMENT

3.3.1 Environmental management

From a western, scientific viewpoint, environmental concern and conservation was coming of age in the middle of this century with landmark publications such as Leopold's (1949) *A Sand County Almanac* and Dasmann's (1959, 1984) *Environmental Conservation*. The increasing industrial output and efficiency led to pollution awareness. Despite early warnings of the dangers (Wigglesworth, 1976), the realm of entomology was partly responsible, with a rapidly increasing use of organic pesticides in the 'the only good bug is a dead one' era of pest control in the 1950s and 1960s. Professional and public opinion was then alerted to the general adverse impacts that these methods of reducing unwanted insects were having on fauna and ecosystems ecologically and geographically removed from the application areas (Carson, 1963; Mellanby, 1967; Moore 1987a). Other entomologists and conservationists were fully aware of the significance of insect conservation, particularly within the context of preservation and management of their habitats and the ecosystem (Duffey and Watt, 1971; Morris, 1971; Duffey, 1974; Moore, 1987a).

3.3.2 Gaia

The closing of the twentieth century is seeing a distinctly strong swing from anthropocentric to holistic thinking. We are connected with, and are part of, all around us. Davies and Koch (1991) even argue that virtually electrons and nuclei of the atoms that are or have been part of living matter on earth came from almost all stars in our nearby galaxies and even from all other galaxies in the universe.

The recent sharp jolt concerning the destruction of our biotic world (Wilson, 1985) at ground level also coincided with the clear awareness of the interrelatedness and interaction not only between living organisms, but also between them and their atmospheric shell. Earth functions rather like an organism itself: 'Gaia' (Lovelock, 1979, 1988). According to the Gaia hypothesis, organisms not only interact among themselves and have their evolutionary paths influenced in turn by abiotic environmental changes, but the organisms, as a whole, influence the atmospheric environment. Hoyle's (1948) famous and prophetic quote 'Once a photograph of the Earth, taken from *the outside*, is available.... a new idea as powerful as any in history will be let loose' did indeed generate a renewed reverence for all life, and on this unique earthly home, insects are the genetic majority.

3.3.3 Undoing the damage

Today some consider that wild nature has come to an end completely. There is no place on earth where man's exhaust gases do not in some way taint the world (McKibben, 1990). Global repair will have to come through international agreements aimed at slowing, if not halting, the harmful global impacts. The problem is that the enhanced global effects have momentum, and even stopping emissions will still continue to influence ecosystems. At literally grass-roots level, the new science of restoration ecology, the local repair of damaged landscapes, is beginning to emerge. (Buckley, 1989; Davis, 1989; Jordan, Gilpin and Aber, 1987), alongside ecological landscaping (Bradshaw, Goode and Thorp, 1986). Inevitably, insects, being numerous, genetically diverse ectotherms, will benefit from any restoration activities, whether local or global. The species to benefit most will be the eurytopic ones with wide geographical ranges. These topics are further covered in Chapter 11.

3.4 PROBLEMS OF PROTECTION: FROM SPECIES TO ECOSYSTEMS

3.4.1 Species protection

Known threatened species are listed in *The 1990 IUCN Red List of Threatened Animals* (IUCN, 1990), various national Red Data Books and *The IUCN Invertebrate Red Data Book* (Wells, Pyle and Collins, 1983). In a broader ecological context, there has been concern for conservation of entire ecosystems or, in the case of Antarctica, recent international consent that the whole continent should be conserved. Conservation of these larger areas in their entirety is the most accepted approach to conserving insect species and their habitats.

Very few insects, even many of those on the *Red List,* are individually legally protected species. Many countries do have legislation protecting certain threatened, more conspicuous, often geographically restricted species (e.g. see Heath, 1981; Wells, Pyle and Collins, 1983; Shirt, 1987). Some countries do nevertheless have blanket protection, at least in terms of collecting, of whole insect groups. All Odonata in Germany are 'protected' from collectors but not necessarily from hyperacidic rain which may be more damaging.

In practice, most insects are protected by preservation of their biotopes and ecosystems, especially those that are already reserves of a general nature, whether of special scientific interest or of scenic beauty.

Collecting of species, even rare butterflies, by enthusiasts seems to have had little impact on insect populations and not to have caused any extinctions as a singular impact (Pyle, Bentzien and Opler, 1981). This is not to suggest complacency. Certain species are highly prized by collectors because of their rarity, size and beauty (Morris *et al.*, 1991). This 'specialist collecting' could pose a threat over and above biotope modification (New and Collins, 1991). In Europe, such species include the Corsican swallowtail (*Papilio hospiton*), while in Asia there are many much sought-after species. One is the Kaiser-I-Hind swallowtail butterfly (*Teinopalpus imperialis*) of the Himalayas, which is collected by foreign entomologists in Nepal waiting for hill-topping males (New and Collins, 1991).

In some countries, there are specific codes laid down for insect collecting, the most well-known being 'A Code for Insect Collecting' established by the Joint Committee for the Conservation of British Insects under the auspices of the Royal Entomological Society. This code is reproduced in Fry and Lonsdale (1991).

There are international conventions that protect species. The 'Bonn Convention' is the Convention on the Conservation of Migratory Species of Wild Animals, 1979. The 'Berne Convention', Convention on the Conservation of European Wildlife and Natural Habitats, 1979, outlaws the collection or possession of listed species. The Berne Convention is of particular relevance to insects because it is biotope orientated. In Britain, the 1981 Wildlife and Countryside Act is also highly beneficial to insects in that it provides protection of biotopes in localities designated Sites of Special Scientific Interest, the selection of which takes into account the habitats of rare insects.

The interface between conservation of individual species and the relative significance of larger land units has been addressed by the analysis by Collins and Morris (1985) of critical faunas. As conservation management implies use of human resources, it has a political element and therefore involves political boundaries. This overlays the natural distribution range and distribution pattern of species and of ecosystems.

3.4.2 Biotopes, reserves and national parks

Agriculture and urbanization have modified the fertile, low-lying, well-watered ecosystems, with only fragments being left for conservation of biota. There are some exceptions: the 1 900 000 ha of the Kruger National Park in South Africa is of great agricultural and mining potential, but fortunately for the insects the area is rich in large mammals. Even so, populations of these mammals are managed to keep the landscape as near pristine as possible, without overgrazing, which, in turn, benefits the natural insect community. But in critical appraisal, the Kruger National Park for example, although large and an important island refuge for many taxa, is not of great value in the sense of being rich in locally endemic species. Its high value lies in its size coupled with its typicalness of a particular biome, the southern African savanna. It is not an endemism epicentre, but is a large, relatively undisturbed tract of land representing a once much more widespread landscape.

Most of the earlier established large reserves and national parks were mostly chosen fortuitously for their combination of scenic splendour and poor agricultural potential, with mountain or canyon-inhabiting fauna of all sizes receiving automatic protection. In complete contrast, areas such as the fertile prairies of North America have not only seen huge declines of many vertebrates, but are hardly pristine anywhere, with a concurrent disturbed insect fauna (Murphy, 1989).

Political factors are also often behind the original choice of nature parks, and, in the words of Kingdon (1990), once an arbitrary block of territory has been institutionalized into a national park, it quickly takes on in the public's mind and speech an identity not less distinct than a great city or mountain.

Insects are not limited by the game fences, but they are by their habitat – those on the inside continuing to survive as they always have done, and those on the outside being the opportunists able to survive the modified landscape, be it grazed by cattle or planted to a crop.

To date, nature reserves occupy four million square kilometres of land,

and in addition there is an unknown amount of wilderness under private or provincial protection (Durrell, 1986). Even if this land were properly managed, and even if the figure trebled by the turn of the century, which appears quite untenable given the human population pressure, only about 10% of the world's land surface would conserve natural areas.

If we set this against the basic tenets of island biogeography theory (MacArthur and Wilson, 1967), according to which, in approximate terms, a 10-fold decrease in area results in a halving of the number of species, we can expect thousands of species of insects to become extinct within the next few years. By the year 2000, several thousands of species extinctions per year could be expected. This is mainly the result of habitat loss and landscape modification. This figure will only ever be an estimation, because most of the species will never be known to science. The figure is likely to escalate in keeping with global climate changes, especially in years of adversity, such as exceptionally hot or cold, dry or wet years. These crucial years are likely to filter out species as natural selection is subject to the harshest conditions, not the average ones.

3.4.3 'Biosphere reserves' and multiple-use modules

Conservation cannot favour simply the biota without regard for human communities. Combining the needs and aspirations of local human communities with nature reserves is the principle behind the UNESCO (United Nations Educational, Scientific and Cultural Organization) 'Man and Biosphere' (MAB) programme. This is a broad approach to protecting ecosystems in harmony with man. A 'biosphere reserve' has a core of true natural wilderness, large enough for both plants and animals to continue at their natural levels (Figure 3.5). This core is surrounded by one or more buffer zones. In these buffers, there is human activity from light resource utilization such as some wood gathering and hunting, with established, and principally traditional, human settlements. The reserve is run on an open and integrated dialogue approach, with decisions being made cooperatively between scientists, local people and managers.

The biosphere reserve programme covers several areas in each of the 190 or so biogeographic provinces, with the aim of nature protection coupled with sustainable utilization. There have been practical problems in implementing such a massive plan. From an insect conservation biology point of view, such reserves are valuable in view of their size, and variable biotopes, including ecotones.

One of the problems with the concept of a biosphere reserve is that it is an island, albeit a big one in many cases. Harris(1984) and Noss and Harris (1986) point out that real ecological processes and activities function in a time–space mosaic across a full hierarchy of measurement. They emphasize

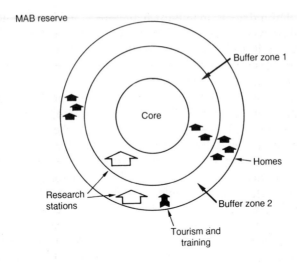

Figure 3.5 A 'Man and Biosphere' (MAB) reserve illustrating how both man and biota benefit by buffer zones where sustainable utilization of resources can take place. Nevertheless, there are still threats to many of these reserves from increasing human pressure.

·that conservation should consider heterogeneous landscapes, which allows optimal ecological patterns and processes to take place. Noss and Harris (1986) propose that area nodes of high ecological value be connected into a functional network. These networks would preserve diversity at all scales. The nodes could be multiple-use modules (MUMS) which consider human needs and activities as well as those of wildlife. A MUM network would protect and buffer important ecological entities and phenomena, while encouraging movement of individuals, species, nutrients, energy, and even habitat patches across time and space.

3.5 NEO-MALTHUSIAN AND ANTI-MALTHUSIAN VIEWPOINTS

Malthus (1798) pointed out that the human increase could not continue indefinitely, and that 'misery and vice' must eventually limit the population level. There are today two opposing viewpoints as to the cause of our environmental problems. The neo-Malthusians blame the developing world's ills mostly on population growth. The anti-Malthusians, in contrast, blame factors such as inappropriate technology, overconsumption by the affluent and inequality and exploitation, which force poor farmers into marginal land which inevitably becomes degraded. The truth probably lies somewhere in between, with three key factors (Harrison, 1990). The first is the level of consumption, determined by lifestyles and incomes. The second factor is the technology needed to satisfy that consumption, and to dispose of the waste generated. These two factors interact to give the amount of environmental damage done per person. This is then multiplied by the third factor, human population, giving the total level of damage.

For insect conservation there is an irony in this debate. Increased urbanized standard of living and education go roughly hand in hand. Information flow through television adds to these. But the increased standard of living generates waste and pollution, yet the population becomes environmentally aware, often too late. With this environmental awareness comes a much greater appreciation of small animals such as insects which are magnified in colour plates and on the screen. This is not to say that all human societies are out of tune with insect behaviour and ecology. Many human communities that still live in harmony with their natural surroundings revere certain insects.

There is a greater ratio of people interested in insect life relative to faunal richness in temperate lands than in the tropics. The reason for this is twofold. There are fewer species in the temperate lands and also the average standard of living is higher, with associated greater media exposure and nature information flow. In Britain, the human population is about stable, and there are 45 species of dragonfly. There is an active British Dragonfly Society with over 1000 members and a strong conservation arm. This gives over 22 committed enthusiasts per species. In insect-rich Venezuela there are at least 449 dragonfly species described to date and no national society.

Across the world, insect conservation concern is generally inversely related to human population growth rate, overall education standards and gross national product (GNP), with the exception of some human communities and sectors in many of these countries. Nevertheless, certain in-depth autecological studies that have taken place in temperate lands (e.g. J.A. Thomas, 1991) provide valuable insights for insect conservation at all latitudes. For example, it is important to recognize the different vulnerabilities of the various life stages of insects. But there must always be caution in extrapolation from one species to another, and particularly from one insect community to another.

3.6 HISTORICAL REVERENCE FOR INSECTS

3.6.1 Glimpses of ancient viewpoints

Many ancient cultures, more in harmony with nature than we are today, revered insects. Some species, such as the scarab beetles (genera *Scarabaeus*, *Copris* and *Catharsuis*), were, to the ancient Egyptians, highly symbolic that all life springs from the sun. The Japanese have delighted in dragonflies for centuries, and even have beautiful folk songs referring to them. Such cultural associations run deep, and several Japanese reserves are devoted specifically to dragonflies (Moore, 1987b). For the first of these reserves there is even a fully illustrated handbook devoted to identification of the species (Sugimura, 1989). The North American Cherokee Indians were fond of fables, and with a poignant message for today's biodiversity crisis, one story tells of a hunter who was surprised and killed by the enemy because he sneered at the song of the katydid.

3.6.2 Medieval outlook

In medieval England, insects were very much part of everyday life. Body lice were even called 'pearls of God', being considered a sign of saintliness. As the body of Archbishop Thomas à Becket lay in Canterbury Cathedral on the night of his murder, the cooling of the body caused the insect parasites to crawl out from beneath his robes. It is said that this epizoic fauna 'boiled over like water simmering in a cauldron, and the onlookers burst into alternate fits of weeping and laughter, between the sorrow of having lost such a head, and the joy of having found such a saint'.

Apart from devastating plagues of the Middle Ages vectored by certain fleas (*Xenopsylla* spp. and *Pulex irritans*), there was generally great reverence for insects right into the twentieth century. Of course the epidemiology and the exact role of fleas in the disease was not known at the time, but the effects were certainly severe. In the year 1349, some 50 000 people died in London alone. Construction of an immense cathedral at Siena, Italy, begun in 1339, was interrupted by the plague of 1348 and never resumed. The present great cathedral was to have been merely its transept. Only the façade and the bases of the pillars of the nave remain of the original vast enterprise.

3.6.3 Nineteenth century

The nineteenth century was the grand age of entomological curiosity and romanticism. Insects, by their diversity and conspicuousness with the awakening of European springtime, became wondrous creatures in the tapestried scheme of nature. Insects were marvelled at and eulogized. The three-volume *Episodes of Insect Life* authored by the hearth cricket (*Acheta domestica*, 1849), refers to entomology itself:

> ...Let us proceed to put her through her paces, and show how in pursuit of 'charming variety' she carries us through roads as varied. Now like an ambling palfrey, she bears us over flowery meadows; now, like a flying Pegasus, mounts with us through air; now descends beneath

earth's surface; then plunging in the stream, opens to us new worlds beneath the waters.

Reverence for nature, including a strong awareness of invertebrates, is returning, albeit through the sudden magnitude of species range retractions and extinctions. The new approach is not so wordy, but scientifically crisp, alarming and desperately urgent.

3.7 EMERGENCE OF THE SCIENCE OF CONSERVATION BIOLOGY

3.7.1 Shift from local to global concern

It is the invertebrates that are 'the little things that run the world' (Wilson, 1987). Invertebrates as a whole are more important in the maintenance of ecosystems than are vertebrates, and *in situ* conservation through simple habitat protection, with or without minimal management, is viable and cost-effective.

At the close of the twentieth century, perspectives on conservation have changed. In North America, national parks were initially principally preserved for their scenery, and in Africa for large vertebrates. In Europe, the starting point was different. There was a strong move to preserve the minute remnants of ancient woodland, fen and heath, as well as centuries-old already-disturbed landscapes such as chalk grasslands and hedges (Moore, 1987a). Today all the land available for reserves is virtually taken, and at best there must be much greater sustainable integration of man and the environment, to stop the demise of more insects.

Also, as this century closes, conservation action has broadened from local to global concern, from vertebrates to invertebrates, and from isolated large vertebrate species to the interactiveness of whole ecosystems. Green activists and professional conservationists are hearing the same tune, as have various of the world's religions (Anonymous, 1986, 1990).

3.7.2 Definition of conservation biology

Soulé and Wilcox (1980) described conservation biology as a mission-orientated discipline comprising both pure and applied science. Soulé (1986) even called his edited book *Conservation Biology: The Science of Scarcity and Diversity*. Murphy (1990) gives a concise definition of conservation biology – 'the application of classic scientific methodology to the conservation of biological diversity'.

The surge of interest in conservation biology has coincided with much greater awareness of the human population growth, global impacts and loss of so many natural biotopes. But for some time there has been great awareness of the significance of the scientific basis of conservation of organisms within their particular habitats and in the landscape. Among the earlier works that covered scientific aspects of conservation of wildlife both in nature reserves and in surrounding areas include those of Ehrenfeld (1970), Duffey (1971) and Duffey and Watt (1971). The gap in the intervening years to the present has prompted little interest by pure ecologists. Practically minded conservation biologists have been extremely small in numbers, particularly those working on invertebrates.

Earlier terms included 'conservation ecology' and 'biological conservation'. The real difference between these terms and the latest, 'conservation biology', is that this new term was launched to emphasize a crisis science recognizing the immediate and major adverse impacts facing our biosphere. In turn, 'insect conservation biology' is simply referring to the major faunistic component of biotopes, ecosystems, landscapes and biomes under threat from both local and large-scale effects.

3.7.3 Practical value of insect and plant diversity for insect control

Aside from the important ethical views on insect conservation (Chapter 9), there is the possibility of a Darwinian acceptance that, as 99.5% of species through time have already become extinct through natural causes, man is simply a new form of intensive, rapid and instantaneous form of selection pressure. The difference now is that we, you and me, are causing these extinctions 1000–10 000 times faster than non-human nature (Wilson, 1988). A rough estimate is that between the years 1990 and 2020 species extinctions caused primarily by tropical deforestation may eliminate between 5% and 15% of the world's species. Based on the estimate that there are about 10 million species on earth, this would amount to a potential loss of 15 000–50 000 species per year or 50 to 150 species per day (Reid and Miller, 1989).

Pragmatically, too, we must keep our options alive (Reid and Miller, 1989). Where might lie an economically valuable future biocontrol agent, like *Rodolia cardinalis* from Australia, which saved the Californian citrus industry in 1889 from the ravages of the cottony cushion scale (*Icerya purchasi*)? Or where might we find another weevil like *Elaeidobius kamerunicus* which has replaced hand pollination of oil palm to the value of US $13 million per year (CIBC, 1982–83)? Where might a new gene lie for incorporation into a biocontrol agent to make it more heat tolerant?

On the other side of the coin, conservation of plant diversity also means the possibility of finding new, natural insecticides (Prescott-Allen and Prescott-Allen, 1982; Plotkin, 1988). This is not a pipe dream. Applied entomologists are well aware that natural products such as nicotine (e.g. against several glasshouse pests), pyrethrum (e.g. against various pests, including insects on animal pets), rotenone (e.g. against garden pests) and sabadilla (e.g. against thrips) have been used as insecticides for years (rotenone since 1848 against leaf-eating caterpillars) and have low mammalian toxicity. Much research continues and with synthetic pesticides now costing over $100 million to launch, and with chemical resistance becoming so widespread and additive upon environmental repercussions, the search for new suppressive, environmentally safe, plant-produced insecticides is an important field of development.

3.8 THE BOTTOM LINE: WHO PAYS FOR INSECT CONSERVATION

The rosy periwinkle (*Catharanthus roseus*) is a small, attractive flower that grows in the forests of Madagascar. It has been in the limelight because a drug extracted from it is the only known cure for a certain kind of childhood leukaemia. This is one of many medicinal plants from the rain forests that are of benefit to mankind.

Schiøtz (1989) presents a real, bottom-line scenario that applies not only to the rosy periwinkle but also to, say, insect biocontrol agents and insect pathogens, areas of pragmatic biodiversity conservation that benefit mankind in general. Schiøtz (1989) argues along these lines. Why should Madagascar preserve the rosy periwinkle? The world community benefits from the plant, but what are the benefits to the local people or the Madagascan government? The answer is: None at all. No money flows back to Madagascar for the drugs produced, and it is even unlikely that the drug itself is available to the poor peasants of Madagascar.

Although major initiatives have come from major non-governmental organizations such as the World Wide Fund for Nature, the reality is that the poorer countries are caught in the loop of high birth rates, coupled with increasing aspirations for a higher standard of living, yet increasingly greater pressure on resources, which in turn causes greater environmental disturbance to the detriment of biodiversity. Now that the east–west military conflict has abated, we must mobilize these enormous resources, although not in military hardware terms, but in sustainable development programmes, to resolve this impending north–south conflict, for the sake of man and all other biota. Whatever our principles, ethics or beliefs, the biodiversity crisis and this enormous loss of insect life is happening now, a time when economics is the principal dictator.

3.9 SUMMARY

In some north-temperate countries, particularly Britain, there has been focus on individual high-profile species, of which only the subspecies may be endemic. This autecological approach has not neglected conservation of the biotopes and ecosystems, albeit in a historically fragmented landscape. But in terms of overall genetic loss, it is the tropics that are demanding saving of landscapes as an umbrella for the vast array of largely unidentified insect species. Not that the tropics are without flagship species, with the magnificent Queen Alexandra's birdwing butterfly highlighting a conservation area.

Insect species decline and loss has given emphasis to environmental management, particularly as there are so many species that individually or collectively act as environmental barometers. Awareness of the magnitude of biodiversity, and its impending rapid loss, has coincided with the concept of a holistic earth, Gaia. Wild nature has gone, and restoration activities are now under way. Insects to date, despite their fascination, variety and major role in ecosystems, generally have been given little attention.

Insect awareness within the framework of biodiversity loss has started to gain momentum. Insects are beginning to feature strongly alongside plants and vertebrates on lists of threatened biota. They are also beginning to be covered by international conventions protecting species through habitat and landscape protection and restrictions on trade in threatened species.

Reserves and national parks, although not necessarily designed for insect protection, play a major role. But man and his needs must be built into the equation, and reserves such as 'biosphere reserves', which have a range of levels of disturbance decreasing to a core, may be particularly significant for insect diversity protection. Hand in hand with this is the necessary emphasis on educating the increasing human population that insects are important to man.

Insects in past centuries were revered. Then, with increasingly intensive agriculture and an awareness of their pathogen vectoring behaviour, they fell into general disrepute. Their conservation has depended on specialists not only distingushing between 'good' and 'bad' insects but also recognizing that insects are indeed important components of the biosphere.

With the emergence of an increasingly scientific approach to biodiversity conservation, insect conservation has in recent times started to receive critical appraisal. Conservation biology is a crisis science. As there is such an enormous taxonomic impediment with insects, resourceful, urgent and applicable science is required. It is vital that entomologists emphasize that economically exceedingly valuable insect species may await discovery. Saving as many biotopes and landscapes as possible is vital. Diversion of funds from wealthy nations to poorer ones is urgently required.

Part Two
Levels of Analysis

4

Scaling and large-scale issues

Most of us have lost that sense of unity of biosphere
and humanity which would bind and reassure us all
with an affirmation of beauty.

Gregory Bateson

4.1 PROTECTION OF INSECTS AND WHERE THEY LIVE

4.1.1 Habitat destruction and biotope modification

The term 'habitat' does not have the same meaning as 'biotope'. 'Habitat' is an autecological concept emphasizing the interaction between the species and physical habitat structure. 'Biotope' refers to the physical local area where a species or several species live.

The protection of certain insects and their biotopes implies the setting aside (with or without management) of areas of land supporting, within these geographical limits, a certain assemblage (all of one taxon, e.g. dragonflies) or community (of various taxonomic groups, e.g. all locally interactive biota).

Conservation of insect species and their habitats (not the singular 'habitat' when 'species' is plural) refers to each species in turn with their specific life-sustaining requirements. No two species have identical habitats. In the case of insects, conservation implies conservation of their habitats, as well as the insects themselves (Collins and Thomas, 1991).

Biotopes can be modified physically, but habitats cannot (although 'habitat structure' can). Biotopes can be modified but not destroyed. Habitat destruction means disturbance of the biotope such that a particular species is unable to survive any longer. In other words an insect can lose its habitat.

4.1.2 Habitat structure

Habitat structure has three main axes:

1. 'Heterogeneity', which encompasses variation attributable to the **relative** abundance (per unit area or per unit volume) of different structural components;
2. 'Complexity', which covers variation attributable to the **absolute** abundance (per unit area or per unit volume) of individual structural components; and
3. 'Scale' which refers to the variation attributable to the size of the area or volume used to measure heterogeneity and complexity (McCoy and Bell, 1991; McCoy, Bell and Mushinsky, 1991).

'Habitat' brings in a fourth axis, which covers ecological and evolutionary facets (e.g. Moran and Southwood, 1982) and which, for insects, refers also to the insect's behavioural ecology, the developmental stage and its life-history style (Figure 4.1).

Habitat is interactive between organism and habitat structure, and the strength of this interaction continually shifts. Further, this shifting has a genetic base. In traditional terminology, there is, in insects, genetic variation for habitat selection (Jaenike and Holt, 1991). This is a behavioural axiom

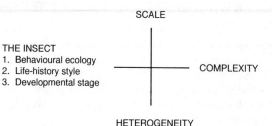

Figure 4.1 A graphical model of the components of habitat, including ecological and evolutionary aspects of the insect. The three components, scale, complexity and heterogeneity, represent habitat structure (McCoy and Bell,1991). They interact with the insect to make up the habitat.

emphasizing the inherent variation in the fourth axis in addition to the other three representing habitat structure. In other words, habitat is a genetically variable concept, as is niche, and inevitably will change with time as physical and biotic conditions change.

Terrestrial habitat structure for insects, which primarily means plant architecture, has varying degrees of complexity. Generally, the more complex the plant architecture, the more herbivorous insects are supported (e.g. Lawton, 1983). The influence of insects and other herbivores can marginally or greatly influence plant form as meristematic tissue is damaged (Mopper *et al.*, 1991). In the case of palms, extensive damage to the apical meristem may readily kill the whole plant.

Insect herbivores may be beneficial as well as harmful to plant growth and reproduction. As well as enhancing or diminishing their own plant resources, insect herbivores modify plant architecture, which in turn influences the rest of the insect community (Mopper *et al.*, 1991). Repercussions extend from the organismal level through population to the community level. Over time, during plant secondary succession, where there is a two-way interaction between insects and plants, habitat structure changes in both heterogeneity and complexity (V.K. Brown, 1991). This continual feedback process is influenced additionally by the physical influence of environmental variables and by feedback loops with parasitoids, predators, mutualists and pathogens.

4.1.3 The plantscape

With few exceptions (e.g. Dempster, 1975; Samways, 1976, 1977a; Denno and Roderick, 1991; Thomas, 1991), we know very little of the behavioural interactions between insect behaviour and habitat structure. In other words, although we frequently refer to conservation of 'insects and their habitats', we use the terms loosely. Among all the insects, we have little information on the relationship between any one developmental stage and plant architecture. Further, there are the different behavioural ecologies and mutual interactions between the various life stages and plant architecture also to consider. Of particular note is the developmental polymorphism in endopterygotes. Lepidoptera are exemplary, being numerous as individuals and species, and by the larvae and adults usually having different feeding habits. Also, extrapolating from other animal groups, it is likely that habitat, including its behavioural and demographic component, will vary across the landscape (Pulliam and Danielson, 1991).

As conservation is a practical scientific management method, it is conceptually easier to refer to the 'plantscape' rather than habitat structure. This is especially so for insects, as the plant–insect interaction is the dominant biotic interaction. 'Plantscape' all-embracingly covers architecture of individual plants and plant communities, and the spatial relationship between the plant forms. The plantscape is successional, and also amenable to management (e.g. Morris, 1991). It is a smaller scale component and has compositional and management comparisons with the landscape discussed in Chapters 5 and 6. 'Plantscape' assumes no knowledge (as 'habitat' does) of insects' behaviours and evolution.

4.1.4 The plantscape and insect behaviour

The influence of the plantscape on insect behaviour is well illustrated by the responses of bush crickets in southern France to the vegetational components and architecture (Figure 4.2).

Platycleis intermedia is susceptible to vertebrate predation on the ground, so it climbs into a bush at night to broadcast its proclamation song (Samways, 1977a). This is a spatial phenomenon relative to plant height and type of complexity. Within this plantscape, microclimatic conditions change temporally as well as spatially which naturally has physiological consequences as it does for other ectotherms (Huey, 1991). Removal of the bushes causes the bush cricket to disperse by flying to a suitable position nearby in the same biotope. This is the insect's response to a particular change in plantscape. But its habitat remains unchanged i.e. its behavioural ecology and its life-history style, and the heterogeneity and complexity of the plant architecture that it inhabits at this scale, are unchanged, .

In the same biotope in France, another bush cricket, *Tettigonia viridissima*, prefers tree rather than bush architecture in the plantscape. It too has a circadian pattern of movement that has behavioural selective advantages. Changing of the plantscape, say by removal of the tree, again does not change its habitat, but does destroy it.

The plantscape may be defined without reference to the insects or other organisms that the plants support. The plantscape is simply the structural and

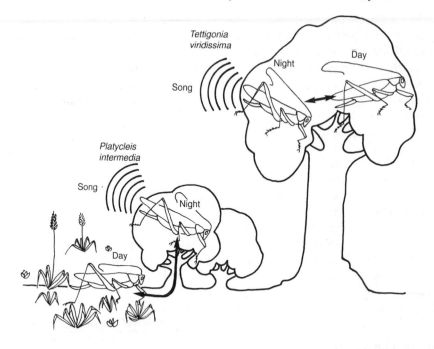

Figure 4.2 The plantscape, the spatial and structural arrangement of plants, is an important concept in practical insect conservation biology. Insects do not always remain in the same place throughout a 24-h period. Because of differential movements of various species of insects and the varying composition of plant species, the 'habitat' is not easily definable. Here is a stylized depiction of a plantscape in southern France, home to the tettigoniid bush cricket *Platycleis intermedia*. During the day it rests among grasses and forbs near the base of bushes, into which it ascends at night to broadcast its calling song. In the same locality, another species of bush cricket, *Tettigonia viridissima*, inhabits trees. This bush cricket moves to the periphery of the canopy at night to sing, but shelters deeper within the crown during the day. Only the males sing, and females (illustrated here) locate the singing males. These three-dimensional aspects are extremely important for insect behaviour and conservation.

spatial composition of the vegetation. In short, the plantscape provides the vegetational foundation for many overlapping habitats. To emphasize that the plantscape is important for other animals besides insects, Uetz (1991) reviews the influence of plant architecture and spatial pattern on spider behaviour.

4.1.5 Microsites

Microclimatic differences may vary enormously from one part of the plant to another, even well away from the microclimatic influence of the soil

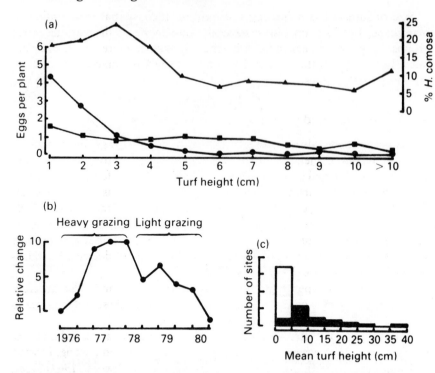

Figure 4.3 Requirements of the Adonis blue butterfly *Lysandra bellargus* in Britain.
(a) Distribution of *L. bellargus* eggs ●, *L. coridon* eggs ■, and its foodplant
Hippocrepos comosa ▲, in turf of varying heights. (b) Relative changes in
numbers on a site that was heavily grazed in 1976–78 and lightly grazed in
1978–80 compared with sites under constant management. (c) Mean height of turf
on 64 downs (grassland hills) containing abundant *H. comosa* in 1978: (☐)*L.
bellargus* present; ■, *L. bellargus* extinct. (From J.A. Thomas, 1991.)

surface. This has been shown to have conservation implications for larvae of
British butterflies (J.A. Thomas, 1991). As climate changes with season, there
may also be seasonal shifts. The minute temperature-sensitive parasitoids
Aphytis spp. show such sensitivity. At sites in the lowveld of South Africa,
both *A. africanus* and *A. melinus* population abundances change balance in
the tops and bottoms of citrus trees depending on the seasonal changes in
humidity and temperature (Samways, 1985).

   The significance of the structural component of the habitat for animal
species survival has been known for some time (Elton, 1966). Its significance
for insect conservation has also been recognized for many years (Morris,
1971; J.A. Thomas, 1991) (Figure 4.3), leading to the recent work on the
structure of the tropical forest in maintaining vast numbers of individuals and
species both on the forest floor and up through to the canopy (Stork, 1988).

Size of insects has a bearing not only on their spatial exploitation of plantscape, but also on our taxonomic knowledge of the various insect groups. Large insects eat more and generally require more space. They are much better known than the small species of the same taxon, especially in the species-rich tropics.

4.1.6. Insect size and plant relationship

The concept of fractals is a valuable way of viewing plant geometry relative to insect sizes and their microsite preferences (Morse, Stork and Lawton, 1988; Williamson and Lawton, 1991). Fractal geometry refers to the fact that with increasing resolution or magnification greater detail is seen. This is not the case with squares, circles, spheres or cubes, which are rare in nature. But with plant architecture and microarchitecture, increasing magnification reveals an increasing number of nooks, crannies and rugosities that provide shelter for insects. A conceptually simple construction to describe fractals is the Koch curve (Figure 4.4).

It is easy to extrapolate how various insect species and their different developmental stages of various sizes can make use of these different scales of detail.

The complexity of the architecture of specific plants has played a role in shaping community structure, particularly guild composition (Strong, Lawton and Southwood, 1984). Tree history also plays a role. Southwood (1961) found with British trees that the older and more widespread and abundant a tree species was in the past, the more species of insect live on it at the present day.

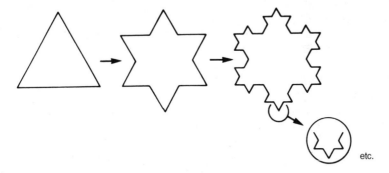

Figure 4.4. Stages in the construction of the Koch curve. At each stage in the creation of this curve a straight line three units long is converted into a kinky line by building an equilateral triangle on the middle third. The resulting stretch of line is now four units long. The construction can be repeated indefinitely, and at each stage the total length of the line becomes four-thirds what it was previously. The line can, in effect, become infinitely long. (From Williamson and Lawton, 1991.)

In terms of spatial scale, the microsite is the basic physical unit where the insect lives. Each insect also has its microhabitat, which is the microsite relative to the insect's behaviour and its responses to the microenvironmental conditions. The microhabitat might be a well-shaded and humid crevice in a tree trunk, and the insect, say a psocid, shows taxis to find these conditions. For example, should a tree naturally fall down and the trunk is now exposed to sunlight, the microsite is unchanged, but the microhabitat has been destroyed and the pscocid either moves to another tree or dies.

As with the larger scale of the biotope and habitat, the different developmental stages of the insect may have very particular requirements to which the animal's relative mobility and life-history style relate. A modified biotope may not adversely affect conditions for the adult, but it may for the larva. Microclimatic conditions may still be suitable for the adult but not for the larva. Invasion of the biotope by a flowering invasive weed such as *Lantana camara* may provide further nectar-providing microsites for the adult

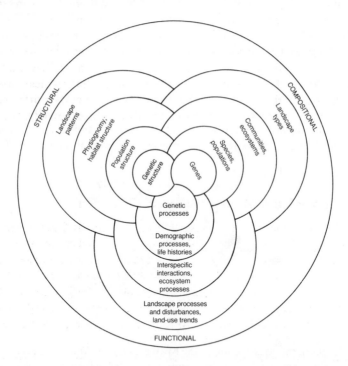

Figure 4.5. Compositional, structural and functional biodiversity, shown as interconnected spheres, each encompassing multiple levels of organization. This conceptual framework may facilitate selection of indicators that represent the many aspects of biodiversity that warrant attention in environmental monitoring and assessment programmes. (From Noss, 1990.)

butterfly, but the plant may shade out the larval food plant at ground level.

4.1.7 Hierarchy of scales: structural, compositional and functional

Biodiversity conservation is an open-ended subject meaning many different things to different specialists. When undertaking a study it is important to be aware of the scale at which one is working. Noss (1990) has conveniently put this into perspective with his principle of hierarchical characterization of biodiversity (Figure 4.5).

He suggests that biodiversity be monitored at multiple levels of organization, and at multiple spatial and temporal scales. All is connected, and no single level of organization (e.g. gene, population, community) is fundamental, and different levels of resolution are appropriate for different questions. Big questions such as global impacts need to be addressed from several scales: structural, functional, compositional and temporal. Before man's impact, generally, the larger the scale, the greater the time process involved, catastrophes aside. Man's impact has changed this approximate proportionate relationship. The global nature of processes is emphasized by the recent suggestion that the British climate is influenced by the warm Agulhas Current way down in the southern hemisphere off the southern east African coast, with a 3-year delay in the transport of heat from the south to the north. Such events impose natural selection pressure on individuals far removed from the original environmental influence.

4.2 GLOBAL MATTERS

4.2.1 The entomologist as conservationist and citizen

Professional entomologists are citizens and consumers, and many are involved in the suppression of pests, while a few are involved in the survival of rare insects. Amateur entomologists also play a major role in the preservation of species, whether by gathering data or by active participation in local projects. Meanwhile, climatologists monitor changes at the global level that will impact on insects. Non-governmental organizations and international agreements continue to play a part in working towards slowing down global pollution and in saving biodiversity.

What can entomologists do towards these ends? Firstly, as citizens and consumers they can take the lead or follow environmentalist issues towards a more sustainable world, from having ecologically landscaped gardens to using environmentally friendly products. But, in particular, they can emphasize to fellow citizens the significance of insects as by far the major component of biodiversity. Some books on biodiversity conservation give disproportionately too little importance to insects, bearing in mind that these animals may make up over 64% of the world's species, and with other arthropods this figure would be about 72%. Insects are important components of ecosystems, and this fact should be emphasized in environmental and conservation biology discussions, as well as in all social and managerial circles.

4.2.2 Experimental studies

Although there has been considerable physiological research on response of insects to temperature, few studies have selected species for greater tolerance to changes in climatic conditions. One of the first attempts to directly genetically select for insects tolerant of heat extremes was by White, DeBach and Garber (1970). Selection was not for conservation reasons but for the biocontrol agent, *Aphytis lingnanensis,* a parasitoid of red scale (*Aonidiella aurantii)*. Selection experiments over 100 generations were partially effective, but more so for lower temperatures than for higher ones.

Parsons (1989) has specifically addressed how enhanced global warming and other effects may influence insect populations, particularly in the tropics compared with temperate regions. A first consideration is that the insects depend directly and indirectly on plants for food and shelter. Parsons' (1989) work on *Drosophila* indicates that there is likely to be extinction and replacement by more tolerant species should rain forests increase in temperature by 2°C. The present enhanced global warming is too fast for directional selection. Within the Australian rain forests, rare species of the subgenus *Hirtodrosophila* generally occur only under the climatically non-stressful conditions of high humidity and a temperature of 18–20°C. As the temperature increases to a limit of about 25°C there is an increasing tendency to find only commoner species at the forest edges and in heat-stressed lowlands (Parsons, 1982). In short, *Drosophila* diversity decreases with increased temperature, with probably a concurrent change in the flora since there is broad tracking by *Drosophila* species of the subgeneric and species floristic characteristics of forests and their resources (Parsons, 1991).

Experimental responses by *Drosophila* to desiccation stress were rapid and substantial, far exceeding other heritable traits such as morphology and life histories (Figure 4.6).

Parsons (1989) points out that these results are in keeping with field observations, but also that there is a metabolic cost involved. The desiccation-tolerant strains had a lower metabolic rate than the unselected sensitive strains, and showed enhanced resistance to a number of generalized stresses, including toxic levels of ethanol, starvation, and intense levels of ^{60}Co -γ irradiation (Hoffmann and Parsons, 1989).

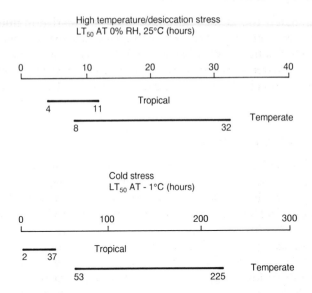

Figure 4.6. Range of LT$_{50}$ values (number of hours it took for 50% of the individuals to die from stress) for nine tropical and 10 temperate-zone *Drosophila* species. For cold tolerance, there is no overlap, and for high temperature/desiccation the overlap is minimal. (From Parsons, 1989.)

4.2.3 Effects in the field

Multiple stresses are the likely scenario for future environmental impacts. Tropical rain forests are at the sensitive forefront, with their general narrow tolerances to change. Peters and Darling (1985) have rightly pointed out that multiple stresses associated with global warming are of primary concern. A steady move away from the adaptive norm of a species is not easy, and in the case of temperature a major player is the enhanced physiological stress. Reactions are strongly canalized to the range of existing populations, and new reactions are never really adaptive (Parsons, 1989). As the environment moves an organism away from the norm, fitness falls as in 'adaptation' of plants to heavy metals (Pitelka, 1988). The point is that insects in general are exceedingly sensitive to temperature and rainfall regimens, and this fact is certainly of value in choosing biocontrol agents, especially in warmer climates (Samways, 1989d). But it is clear that if the enhanced global warming continues, along with other environmental stresses, there is almost certainly going to be another mass extinction, but this time principally of insects, other terrestrial arthropods and plants.

From a converse point of view, global warming may well enhance certain pest insects, or move previously innocuous species into the economic realm to become new pests. Collier *et al.* (1991) have undertaken one of the first experimental studies specifically aimed at determining how global warming may impact on the pest cabbage root fly (*Delia radicum*) in Britain. A rise of 3°C would cause the fly to become active about a month earlier in the season than at present and cause a decrease in synchrony of overwintering populations. The overall rise in temperature would require new strategies for controlling the pest.

More research of the type by Parsons (1989) needs to be undertaken, combining experimental genetic work with field observations. This is needed at different latitudes, across ecosystems, altitudinal gradients and fragmented landscapes. Insects can be valuable animal indicators with their large numbers of individuals, species variations across latitudes (Stevens, 1989), altitudes (Claridge and Singhrao, 1978; Samways, 1989e) and amenability to experimentation, collection and observation in the field.

4.2.4 Loss of specific insect groups

Any estimates at this stage as to the number of species worldwide that may become extinct are really guesses, but it will almost certainly be several thousands, if not many more. There could be as many as 50 000 species per year being lost at the turn of the millennium, with some estimates as high as a 50% loss of all species over the next 30 years. Research on individual species would give an estimate of single species responses, which would be independent of responses of other species in the community. Some species in particular are known to be extremely sensitive. The ground beetle, *Africobatus harpaloides*, of Central Africa is highly sensitive to microhabitat temperature changes which stimulate dormancy when the mean maximum temperature throughout the year varies by only 0.9°C (Greenwood, 1987).

There are few estimates to date on how many species will be lost to enhanced global warming compounded with other stresses, and this may vary in intensity from one area to another. D.D. Murphy predicts that if the earth warms by 3°C, the Great Basin mountains in the western United States will lose 23% of its butterfly species (Cohn, 1989). Henning and Henning (1989) have listed 105 threatened lycaenid butterfly species and subspecies out of a total of 310 species for South Africa. Although some of these are perhaps commoner than supposed, it still does not lessen the global threats of such a locally rich taxon. The listed threats exclude those of increased global warming. Such warming, coupled with the enhanced desiccation and variability of climate predicted for the area, is likely not only to affect the enormously rich flora, but also to have a devastating effect on many of these lycaenids, as well as on other insect groups. Many of these butterflies have highly restricted ranges, some simply in small patches near mountain peaks. Some species may be able to adjust up the mountainside, if time, fragmentation and increasingly restricted space towards peaks allows, but those that frequent the peaks (e.g. *Lepidochrysops outeniqua, Poecilmitis endymion, P. balli*) are almost certainly likely to become extinct (Samways, 1993).

4.3 ECOSYSTEM CHANGES

An increase of only 2°C in overall global temperature would make the planet warmer than at any time in the past 100 000 years (Schneider and Londer, 1984). Inevitably ecosystems will be changed in position and in species composition. Ecosystems have never been in total equilibrium. The apparent constancy of the natural world is an artifact of the temporal and spatial scale we observe (De Angelis and Waterhouse, 1987).

Ecosystems have a large number of components and interactions. The interactions are non-linear and many are adaptable to altered circumstances (Scholes, 1990). Positive and negative feedback loops are widespread, but often involve time-lags.

Ecosystems are dissipative systems in terms of energy, and the system parameters are highly variable in time and space. These factors lead to several generalizations about ecosystem behaviour (Connel and Sousa, 1983; Scholes, 1990):

1. Multiple ecosystems, which may be locally stable, separated by transition thresholds, are much more likely than global stability.
2. Directional change is therefore more likely to be jumpy than smooth.
3. constancy relates mostly to strength and variability of external driving forces.
4. Stability is more likely to be encountered at large spatial scales than small, very short or very long rather than intermediate time scales, and at high integrative levels than low.
5. Environmental predictability is more important than the absolute magnitude of environmental extremes ('harshness') in determining stability and resilience.
6. Resilience is enhanced by previous exposure to stress of the same kind and magnitude; conversely resilience to totally novel stresses is unlikely.
7. Environmental patchiness favours the persistence of particular organisms.
8. There is no simple relationship between diversity and interconnectivity on the one hand, and constancy, stability and resilience on the other.

Scholes (1990) suggests that these points indicate that it is not possible to predict detailed changes in ecosystem change resulting from climatic change. However, terrestrial ecosystems are sensitive to climatic change when key processes such as primary production are tightly coupled to rainfall. In turn, vegetation structure is closely linked to temperature and rainfall seasonality.

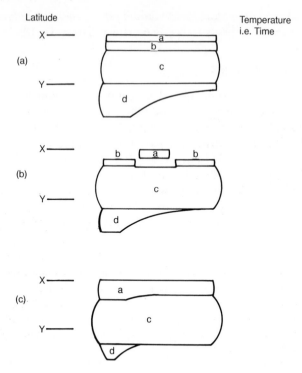

Figure 4.7. Some possible changes in insect species' ranges with increasing global temperatures from now (a), over the next few decades (b), until the end of the next century (c). The scenario is in the southern hemisphere, where endemism is high and species do not have the physiological and behavioural resilience of the species in the north-temperate regions. The species are essentially allopatric/parapatric, and are competitors. As temperature increases (b), 'a' contracts and 'b' fragments, with neither being able to move into the area occupied by 'c'; species 'c' then expands southwards, partially displacing 'd'. As temperature increases further (c), 'b' cannot tolerate the changes and then dies out while 'a' mutates and is able to expand southwards. Species 'a' then does not have the competitive edge over 'c', which expands its range further. Species 'c' partly displaces 'd', which cannot adapt either to the increased competition or the temperature changes. These trends would, in reality, also be influenced by increased degradation and fragmentation of the landscape.

Physiologically sensitive species and immobile ones, particularly if left behind in patches at the retreating ecosystem edge, would be under stress and likely then to become extinct (Figure 4.7).

It is likely that many insect herbivores will follow the fate of their host plants, while these and others, including their parasitoids and predators, will be under their own stress. This will be particularly acute in insects that have

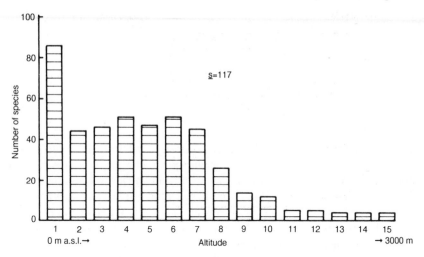

Figure 4.8 Decreasing Odonata species richness with increasing altitude (m a.s.l.) across the 200-km elevational gradient through Natal into Lesotho at about 29°S latitude. Each bar is a 200-m-wide altitudinal segment. The first bar is high species richness from a large tropical component in the fauna, while in contrast the distinctly inhospitable winter climate 2000 m a.s.l. is limiting to all but a few species. (From Samways, 1989e.)

adapted to climatically stable biotopes and are not phenotypically or genetically able to withstand the rapid shifts in atmospheric and soil changes. This immediately highlights the tropics.

It would be valuable to begin long-term monitoring of insects at biome elevational and latitudinal boundaries in selected areas in the world [such work has begun on butterflies in Madagascar (C. Kremen, personal communication) and dragonflies across the 3000-m elevational gradient in Natal, South Africa (Samways, 1989e)] (Figure 4.8).

It would also be valuable to keep records of species abundances at selected reserves, especially if they are patches on the potentially receding edge of shifting biomes.

The sort of data gathered by Wolda (1988, and references therein) at Barro Colorado Island, Panama, and along environmental gradients (Wolda, 1987) will be invaluable for students of the future to make hindsight quantitative comparisons. Even simple 'before' and 'after' checklists as botanists have done for plants (Hoffmann and Cowling, 1990) would be valuable. At present, it is such a race to document what insects are present that local checklists of specific areas will be valuable to future generations of conservation biology investigators. Many of the taxonomists sadly will be virtually palaeontologists, describing lost species like the Antioch katydid, aptly named by D.C. Rentz in 1977 *Neduba extincta* from specimens collected from the now-destroyed sand-dune habitat in California in 1937 (IUCN, 1987).

4.4 EFFECTS ON SPECIFIC ECOSYSTEMS AND BIOTOPES

4.4.1 General considerations

In the past, unless biota was limited by insularity, or barriers like mountain changes and seas, there was a general movement by vegetation belts polewards during interglacials and equatorially during glacials. Prediction of a shift of 300 km is a reasonable estimate based on models and on historical evidence from past warming periods (Furley *et al.*, 1983; Peters and Darling, 1985). Elevationally, for each rise of 0.6°C a species would have to shift up the mountainside by 100 m (Figure 4.8).

This is a theoretical consideration assuming other biotic and abiotic conditions and variables are kept intact and move together. This is unlikely for natural ecosystems. It is even more unlikely when fragmentation of the landscape by man's activities has an inhibitory effect upon population movement and community integrity.

4.4.2 Polar environments

Antarctica, being an island, means that any shift in distribution by biota implies crossing the sea barrier. Additionally, enhanced global warming, thinning of the ozone layer and the increase in snow will possibly be particularly acute. Insects on Antarctica and the subantarctic islands are likely to be severely affected.

However, many low-latitude (northern and southern hemisphere) species have adapted to wide fluctuations in temperature. Those of the Arctic also often occur over relatively wide biogeographical areas compared with their terrestrial relatives in the Antarctic. A further point is that the far northern Palaearctic is still relatively unfragmented, especially in comparison with temperate lands, and many temperate species, having adapted to glacial and interglacial periods, may already possess some potential to move northwards, if there is time relative to the pace of global warming. The tundra species may well be squeezed into tighter belts, particularly in the northern Palaearctic, where the northerly ocean presents a barrier. The fragmentary nature of the many islands of the far northern Nearctic is likely to inhibit movement of flightless insects in particular. Melting of the ice will increase the barrier effect, and rises in sea level (from water expansion as well as melting ice) will enhance insularization.

4.4.3 The north-temperate lands

Many previously widespread north-temperate woodland or forest species with continuous populations are now fragmented into metapopulations or genetically completely isolated populations.

Kellogg and Schware (1981) predict that temperatures will rise by 4–5°C, and most of Europe and the far west of the USA will be wetter, but elsewhere, particularly in the continental centres, areas will be drier. Although it may seem that powerful flying insects may simply move northwards, this depends also on the ability of their nectar sources and foodplants to also move in the time available. This seems unlikely as the upper limit for the spread of temperate trees is about 2 km per year (Bennett, 1986), far slower than tracking the impact of global warming. Additionally, land fragmentation will restrict physical movement, and place species under greater temperature-triggered physiological stress. This stress will be compounded by increased temperatures coupled with other factors such as increased acid precipitation, eutrophication and local pollution, which has been so damaging to the northern European aquatic systems. The green aeshna dragonfly (*Aeshna viridis*) is one casualty of increasing environmental change (Collins and Wells, 1987). Acid precipitation has also been implicated in the decline of other invertebrates such as the noble crayfish (*Astacus astacus*). Molluscs, with their alkaline shells, are also particularly vulnerable to an increase in acidity, which will impact on their insect predators and parasites.

Certain species may be rare and adapted to the centuries-old managed landscape. Changes in these practices, which amount to loss of suitable biotopes, have had a major impact on many insects. Especially notable are the butterflies (Figure 4.9).

4.4.4 The southern hemisphere

There is a great difference in the present status and future prospects facing the south-temperate biomes compared with those of the north (Samways, 1992b). The approximately 22 500 described British species (Shirt, 1987; Fry and Lonsdale, 1991) these probably represent over 99% of the actual extant taxa in the country. This is better than the world total of discovered vertebrates, which is probably near 80%, while probably less than 5% of the world's insects have been described.

In southern Africa, there are 8300 described species of Lepidoptera alone, probably about 80% of the total. The total number of insects in South Africa is about 80 000, possibly only a quarter of the actual total. In Australia, of the more than 108 000 estimated species, only 48 600 have been described (Hill and Michaelis, 1988). Greenslade and New (1991) give a figure of

125 000, which they suggest may be very conservative for the total Australian insect fauna.

The southern tips of the land masses, which are much smaller than those in the north (only one-third of the land surface is in the southern hemisphere),

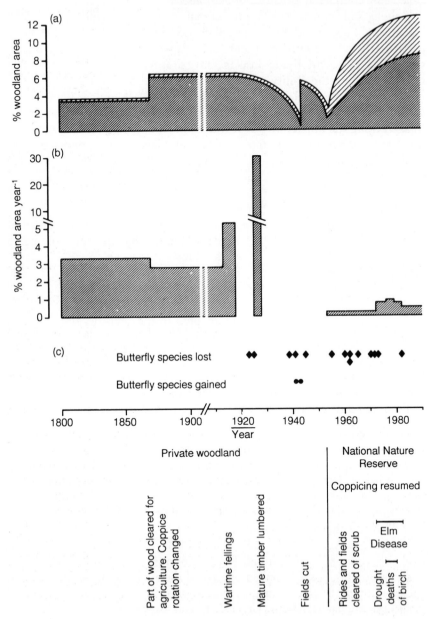

having been virtually free of Pleistocene glacial cover, have species-rich faunas, relatively few taxonomists per species and, of particular note, high levels of endemism compared with the north. Many groups have levels of endemism rivalling the megadiversity areas of the tropics (section 1.5). This makes the southern hemisphere a much more important biodiversity area than is generally realized, with the world's biodiversity pattern being pear-shaped rather than egg-shaped (Platnick, 1991).

These south-temperate species are less likely to be as physiologically tolerant as their north-temperate counterparts as they have not been subject to such climatic adversity and have not followed the retreat of the glaciers.

4.4.5 The tropics

Earlier we saw how important the tropical forests are for insect diversity, with over half of all species living in them. These forests also play a role in stabilizing the atmosphere, and their burning is contributing towards the enhanced global warming. Insects are rapidly losing their habitats and are exposed to landscape and climatic changes. Also, in the tropics insects are adapted only to narrow temperature fluctuations.

There is another factor that is highly relevant to insect conservation biology. Clearing of the trees leaves the red aluminium- and iron-rich soils exposed to heavy tropical rain to be baked by the sun. These tropical soils are infertile compared with their temperate counterparts. They lack phosphorus, potassium, nitrogen and lime. For millions of years the soils have been leached by the heavy tropical rains, and when cleared of long-established multistorey forest, running on the fuel of rapidly eaten and decaying leaves, they bake hard and suit only a few pioneer species. The whole long-established virtually self-sustaining system is broken almost irretrievably, with even the rich soil fauna being severely affected.

From the insect conservation viewpoint, the solution is easy to state but

---

Figure 4.9 Biotope change in Monks Wood National Nature Reserve, England since AD 1800. (a) Permanent open space (rides, tracks, fields), expressed as a percentage of the total woodland area, differentiating between unshaded ground (densely hatched) and shaded ground (lightly hatched, above). (b) Gap creation rate, expressed as the percentage of the total woodland area occupied by new canopy gaps each year. Allowance has been made for retained standards. Gaps were formed by coppicing (to 1914, from 1954), lumbering (1920s), rideside strip cutting (from 1973), Dutch elm disease (1973–82) and drought-induced deaths of birch (1976–79). (c) Last recorded years of butterfly species which appear to have become extinct in the wood, and first recorded dates of colonists. [From Peterken, 1991; historical records summarized by M.D. Hooper in Steele and Welch, 1973; event record, project record and other records on Nature Conservancy Council (NCC) files. Details of estimates have been filed by NCC.]

not so easy to act upon. Quite simply, as much as possible of the forest must be kept intact, with human utilization being kept within sustainable limits. Realistically, it should be recognized that, with the rise in human population growth, keeping the forest intact for the sake of all its biota is vital. Sustainability is not a new concept, and was the way of life for generations of forest dwellers in the past. It is the wholesale, wasteful removal of timber, followed by inappropriate farming methods, coupled with financial motivations and corruption, that has accelerated the present tree removal.

Forest protection services are greatly understaffed, and individuals with a strong moral sense of protection of the forest have been intimidated and even murdered, as happened to Chico Mendes, the Amazonian rubber tapper and union leader who became the world's first ecomartyr in December 1988 when defending the forest against wholesale destruction.

Insect conservation biologists need to provide more data so that non-governmental and governmental organizations can justify proclaiming specific areas worthy of preservation, managed conservation or sustainable utilization. There is a vast amount of research to be done on plant–insect interactions. Fascinating relationships were uncovered by researchers D.H. Janzen and L.E. Gilbert (see Price, 1984 for references) and diversity studies were performed by Erwin (1988) and Stork (1988), biogeographical studies by Descimon (1986) and Holloway (1986) and population variabilities by Wolda (1988).

4.4.6 Coastal areas

Sand dunes and cliffs just beyond the shoreline are often biotopes for rare and localized insect species. The threat to the coast from human population pressure is severe and generally permanent, whether it be from expansion of ports or from ribbon development associated with holiday areas. Additional to this is the impact of enhanced global warming causing an increase in sea level, and the possibility of more frequent inclement conditions buffeting these narrow and fragile landscapes from the seaward side. Ecologically, consideration of these factors must take into account the fact that coastal communities are highly dynamic (Gray, 1991).

During the past few thousand years, the seas have been relatively high and stable, after hundreds of millennia of 142 m variations (Lyman, 1990). In the recent past, the sea level has risen by 10–20 cm per century, but with enhanced global warming the sea level could rise 30-122 cm over the next century, from ocean water expansion, melting of glaciers and possibly some melting of the polar ice cap. A 30-cm rise in sea level would allow seawater to penetrate 32 m inland along a coastline of average steepness, and a 90-cm rise would flood low-lying coastal areas to about 900 m inland.

Such rises in sea level, when coupled with adverse conditions from spring tides to tornadoes, would be detrimental not only to low-lying coastal swamps and to estuaries but also to many oceanic islands, particularly the low-elevation atolls.

Those concerned with insect conservation should monitor species abundances in coastal areas for comparisons by future generations of conservationists. With further confirmation of a rise in sea level, translocation of non-invasive endangered plants (or preservation of their seeds) will be a priority. Many insects will follow as the distances involved would be within many species' home ranges, given time. The complicating factor will be other global impacts, especially the direct effect of warming on the biotas' physiology, which may override conservation efforts in response to the adverse effects of rises in sea level.

4.4.7 Wetlands

Wetland loss worldwide has been enormous, often 90% in parts of the UK, USA and South Africa, as well as many other places. This loss has been mostly to agriculture, although coastal lowlands have suffered urban development, especially around ports.

Insect conservationists are well aware of the impact of loss of wetlands, heralded by the demise of the British large copper butterfly (*Lycaena dispar dispar*). General concern for damage to biota stimulated the formation of the international convention on wetlands protection (The Ramsar Convention) towards worldwide protection of biotically significant wetlands.

Concerning wetlands, there are several implications for insect conservation biology. Firstly, the theory of island biogeography (MacArthur and Wilson, 1967) as applied to nature reserves does not strictly apply. In Natal, South Africa, there has been a loss of about 90% of the wetlands, yet there is no indication that a single species of Odonata or any other insect species has become extinct, yet theoretically about half may or should have done so. There are probably two main reasons. The first is that insects, being small in size and with relatively small home ranges, at least compared with vertebrates, have had their populations fragmented and reduced, but with no harmful impact upon their survival as species. They continue to survive principally in the remnant game reserves, the testing time for which will be the times of drought or other weather calamities, with no mother population to resupply the former restricted populations in the reserves. Secondly, some of the more adaptable species have made use of many farm dams, which although individually small are highly significant as refugia for these animals (Samways, 1989f) (Figure 4.10).

In short, wetlands the world over are under local and global threat. Wetlands are in need of urgent attention from insect conservation biologists worldwide. Many of them are home to rare, localized and taxonomically

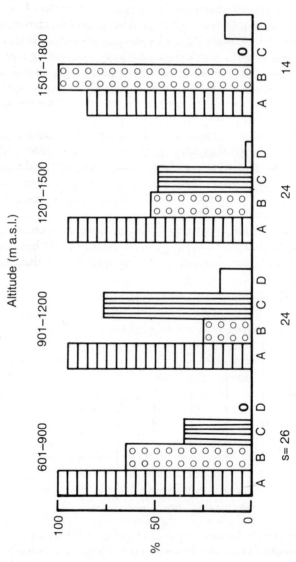

Figure 4.10 Percentage occurrence of Odonata species in artificial and natural water bodies at four altitudes along a 150-km-long transect at about 29°S in Natal, South Africa. (A) Species recorded only from artificial farm dams. (B) Species recorded only from natural biotopes. (C) Species recorded from farm dams but never recorded at the time of sampling from natural biotopes. (D) Species recorded from natural biotopes but never recorded at the time of sampling from farm dams. (From Samways, 1989f.)

isolated species, such as the damselfly *Hemiphlebia mirabilis* of southern Victoria and Tasmania (Sant and New, 1988).

4.4.8 Lakes

Standing water bodies are islands, and at the landscape level track aspects of the theory of island biogeography in that larger water bodies hold more species than smaller ones with the same physiochemical characteristics. This is because there is a wider range of biotopes in the larger lakes. History plays a role with some large lakes, e.g. in Lake Baikal in Russia, having high levels of endemism. As lakes are the focus for disturbance from agricultural, urban, industrial, recreational and transport viewpoints, their fauna is highly susceptible. Eutrophication from agricultural nitrogen run-off is a particularly important community composition modifier.

Yet artificial lakes, or at least ponds, are important for encouraging insect species, particularly the more pioneering types (Samways, 1989f; Moore, 1991a). Some of these have become insect reserves, especially for dragonflies (Sugimura, 1989). Management of such sites is vital, as in sites managed purely horticulturally or aesthetically, and not ecologically, insect species richness declines (Figure 4.11).

As lakes and wetlands are so threatened, it is essential to document species and make suggestions for their conservation, as Foster (1991) has done for the British water beetles. But Reynolds (1991) has pointed out that, as lakes

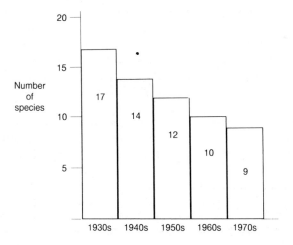

Figure 4.11 Decline over the decades in number of species of Odonata at Richmond Pond, London. The decline has been mostly the result of increased aesthetic landscape manicuring, which has degraded the biotic value of the biotope. (From Fry and Lonsdale, 1991.)

have an island-like quality of available biotopes subject to external threats at a wide range of temporal scales, the biotic components are essentially invasive and adapted to exploit fairly wide levels of variability. The structural organization of the communities is inversely proportional to the frequency of resource opportunities. This means that management is more feasible where variability is less, and an important first step is to determine the regulating factors at each site, and then to determine the conservation objectives, whether for enhanced variability or towards a steady state.

4.4.9 Streams and rivers

Streams and rivers, whether directly or indirectly associated with wetlands, are continually under local threat. This is partly due to their difference from other ecosystems in being vehicles for translocation of materials (Ladle, 1991). Large rivers are known to have lost insect species, such as Tobias' caddisfly (*Hydropsyche tobiasi*), which has not been seen on the River Rhine since the 1920s.

Small streams are often inhabited by rare and localized species. The Mount Donna Buang wingless stonefly (*Rikoperla darlingtoni*) has only been collected within a kilometre of the summit of Mt Donna Buang in Victoria, Australia (Wells, Pyle and Collins, 1983). A necessary and important exercise is simply to undertake more and extensive searches for known rare species, to ascertain their true localness. The fairly recently described damselfly *Chlorolestes draconicus* was known only from one stream in a national park in the Drakensberg Mountains of South Africa. It has recently been found in other streams in a wilderness area elsewhere in the same mountains (Samways, 1992a).

It is vital to monitor insect species, assemblages or communities for several reasons: taxonomic or biotope classifications [e.g. Foster *et al.* (1990) for water beetle assemblages; Wright *et al.* (1984) for rivers]; monitoring pollution impact (e.g. Watson, Arthington and Conrick, 1982); decision-making (e.g. Jackson and Resh, 1989); practical management for conservation (e.g. Sant and New, 1988); and for assessing recovery (e.g. Wallace, 1990). Careful monitoring and comparison of selected sites over the years will help determine global impacts.

Ladle's (1991) description of running waters being a conservationist's nightmare is borne out across the world. The linear and moving nature of the ecosystem means that control has to be along its whole length, as has been achieved with the River Thames in England. But each river has a different course and the biotic communities vary along their lengths, often crossing political boundaries. Thus often there can be no single river conservation policy, with each part of the system demanding particular conservation focus.

4.4.10 Caves

There are many invertebrates, including insects that have evolved to meet the special conditions of their various cave environments throughout the world (Culver, 1986). Many of these species, especially those living deep inside caves, often have reduced pigmentation and long, spindly legs. Physiologically too, after thousands of years of adaptation, their diets are specialized. Although semi-cave species forage outside the cave mouth at night, retreating to the darkness of the cave during the day, it is the permanent cave dwellers that have specialized diets such as bat and bird faeces.

Hawaiian volcanic lava caves contain a range of highly cave-evolved animals with close relatives among the modern, actively speciating native fauna, e.g. the no-eyed big-eyed hunting spider (*Adelocosa anops*) of Hawaii and the thread-legged assassin bug *(Nesidiolestes ana)* (Chapman, 1987). The only representative of the orthopteran family Rhaphidophoridae on the African continent is an endemic and monotypic genus, *Spelaeiacris*, known only from caves on Table Mountain, South Africa (Figure 4.12).

Cave insects occur in isolated populations as determined by their isolated biotopes. Often, these biotopes and their populations are unique. These unique cave conditions are easily disturbed, sometimes even by the slightest

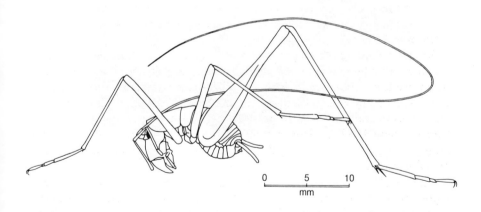

0   5   10
mm

Figure 4.12 The Orthopteran family Rhaphidophoridae is represented in the southern African region only by the endemic and monotypic genus and species *Spelaeiacris tabulae*. (From Ragge, 1974. Reproduced under Government Printer's Authority 9325 dated 7 November 1991.)

Figure 4.13 Batu Caves, in Malaysia, a magnificent unique natural site that is increasingly under great human population pressure through being a major tourist attraction.

interference by man, who, as a tourist or explorer, is intrinsically drawn to caves. The effect may be indirect, with roosting sites for birds and bats disturbed, leading to a reduction in the guano food supply. Where caves are open to the public and easily accessible, pressure from invasive insects, especially the common cockroach, *Periplaneta americana*, adds to the direct human pressure. This has been particularly the case for the famous and enormous Batu Caves, just north of Kuala Lumpur in Malaysia. Originally a Buddhist temple, they are now a major tourist attraction (Figure 4.13).

Caves are a special case for protection. Some, especially in South America, are at present remote and relatively undisturbed. New caves are also being discovered, such as Dragon's Breath Cave, Namibia, which is a single cavity containing a 2-ha underground lake, the largest in the world (Irish, 1989). Several unique invertebrates, including a thysanuran, exist on the floating guano which, because of internal air circulations, is concentrated in one small portion of the lake.

How large-scale effects such as global warming will affect caves is difficult to predict. There will be an impact, due either to changes in percolation of quantities of water into the cave or to a change in vegetation above the cave. Aside from the direct impact of human entry and the invasion of aliens, such as cockroaches, the global effects are likely to be more indirect. Even small impacts are likely to be highly stressful for the cave insects, which have adapted to relatively unvarying environmental conditions over thousands, even millions, of years.

4.4.11 Mountains

For the insect conservation biologist, mountains hold both good and ill for the future. For species inhabiting mountain peaks, global warming may be too severe to allow their survival or the survival of their food plants. The increased ultraviolet radiation due to thinning of the ozone layer may also have highly detrimental effects through increased radiation causing increased metabolic stress on plants. Another disadvantage associated with mountains is that, as orographic regions, they bear the brunt of hyperacidic precipitation from industry. This has been the case in the industrialized northern hemisphere, e.g. the Great Smoky Mountains of the USA. A further consideration is that, as mountains are roughly conical, reserve areas become smaller towards the summits. This means there is less room for movement to suitable areas should environmental conditions change.

One factor in favour of montane species is that many reserves are in scenically spectacular areas that have escaped the plough and the high-rise building. Mountains, being three-dimensional and relatively unfragmented by human landscape activities, are especially important reserves for insects. Insects, provided they are sufficiently mobile, can move around and up and down the mountain in response to daily, seasonal temperature changes. This has been clearly demonstrated with grasshoppers in southern Africa (Samways, 1990b).

Plants on which monophagous and oligophagous insects depend may not be able to withstand the increasing pressure of global environmental stresses, leading to direct insect extinction. However, those plants that are presently at the fringe of their montane range, because of either elevation or aspect

Figure 4.14 The Drakensberg Mountains in southern Africa. This area is a major water catchment area that is managed, principally by fire, to maintain it in its naturally plagioclimax condition. It is an important area for southern African endemic insect species.

limitations, may well be in a new optimal area and flourish at a quicker rate than competitors that were formerly dominant, benefiting some insects but not others. Some lowland species threatened with fragmentation and destruction of their habitats may actually find refuge on the lower slopes of mountains, particularly economically unusable crags and demarcated reserve areas.

A further major consideration is that protection of upland regions means clean, unsilted water for people and biota living further down in the lowlands. This has been acute in the Himalayan region. Minimally managed and almost pristine wilderness areas such as the Drakensberg in South Africa (Figure 4.14), or the rejuvenated forests just north of Kuala Lumpur, Malaysia, are extremely important reserves, especially for insects (Samways, 1989c). They are natural soaks, providing a variety of biotopes for a wide range of insects.

Mountains are clearly an area of hope for many insect species, and other biota. Again, gathering of basic species identification and abundance data now will be invaluable to future generations of conservationists.

4.5 NATURE RESERVES AND GLOBAL WARMING

Global warming puts a new perspective on the concern for nature reserves, managed or not (Morris, 1991). This is especially so in the tropics, where the

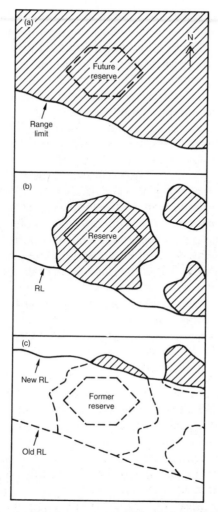

Figure 4.15 How climatic warming may turn biological reserves into former reserves in the northern hemisphere. Hatching indicates: (a) Species distribution before human habitation [range limit (RL) indicates southern limit of a species' range]. (b) Fragmented species' distribution after human habitation. (c) Species' distribution after warming. (From Peters and Darling, 1985.)

genetic and physiological plasticity of species allows little tolerance for adjusting or adapting to changed environmental conditions.

Peters and Darling (1985) have discussed the global impacts on nature reserve design (Figure 4.15). Reserves should be scattered, preferably linked (Harris and Gallagher, 1989), at different elevations and in different areas ready to compensate for global warming. Ideally, a large nature reserve should cover a large area with highly varying topography. But with the fragmentation of the landscape and with little land available for reserves, this is a utopian idea, except perhaps for those species that can survive the local type of fragmentation or disturbance. The grim truth is that we will not be able to forestall extinctions in local reserves under stress conditions, particularly in the tropics. Further, the adverse conditions can come within an ecological, let alone geological, blink of an eyelid. A hailstorm, a frost, a drought, a hurricane or typhoon, or even an epizootic, and a species may be gone.

Pre-emptive translocation could be considered, but for insects, in general, this is out of the question. We know too little of their behaviour and biotope preferences, or of their food plants and nectar sources, for them to be moved from one area to another in anticipation of changes in environmental conditions. Experience to date has indicated that species often do not adapt. Once translocated, even when establishment is initially good, follow-up constant management may well be necessary (Oates and Warren, 1990). A well-documented case is that of the Dutch form of the large copper butterfly (*Lycaena dispar batavus*). This species was introduced into Britain to replace the extinct British form, and constant management has been necessary to maintain the water level for survival of its food plant, great water dock (*Rumex hydrolapathum*) (Duffey, 1968, 1977). But even for this species, management is not so simple, with an interaction between agricultural activities such as peat diggings and cattle grazing.

4.6 INSECT MIGRATIONS AND ROOSTING SITES

The resilient exoskeleton and powerful thoracic muscles have given some insects excellent powers of dispersal, both local and distant, specifically directional and/or determined by wind currents (e.g. Pedgeley, 1982). Directional dispersal may be for resource or mate location, but large-scale mass movement is generally to locate new food resources and oviposition sites. Such migrants are often orthopteroid or lepidopterous pests with known build-up areas and direction of movement, albeit on an irregular basis when weather conditions are suitable for population increase and reduced mortality.

Certain migrations are of great scientific interest. Such migrations include those with distinct flight paths and/or roosting sites. Some insect migrations involve spectacular species that are of no detriment to man's well-being yet are threatened because of pressure on their highly localized roosting sites. The most famous is the monarch butterfly (*Danaus plexippus*)(Figure 4.16).

It overwinters in dense roosts in California (up to about 100 000 individuals at each site) and Mexico (up to 14–25 million individuals at each site). It migrates northwards as far as Canada in the summer (Brower and Malcolm, 1989). Both the Californian and Mexican roosting sites are listed as a

Figure 4.16 Migratory routes of the monarch butterfly, *Danaus plexippus*, in North America. The dots are the overwintering roosting sites, and the hatched areas are the summer distributions. (From Mielke, 1989.)

'threatened Phenomenon' by Wells, Pyle and Collins (1983), owing to pressure from logging and urban development, despite that these roosts are now tourist attractions. The roosts themselves have particular forest canopy characteristics that make them suitable as overwintering sites (Weiss *et al.*, 1991). The permanently occupied sites show a narrow range of indirect radiation from sunlight and a slightly wider but still narrow range of annual direct radiation. Such findings as these have considerable value for management of the forest for protection of this threatened phenomenon.

Of interest is that the disturbed land surface does not appear to have impeded or affected the actual migration paths, as has been the case with many large vertebrates. It is the roosts themselves that are under threat, and not the movement corridors. Further, the species itself is not threatened. Although indigenous to the New World, *D. plexippus* is established in Australasia, Hawaii and elsewhere. Realistic insect conservation biologists must constantly weigh up such paradoxes, which generally may not enter the sphere of vertebrate conservation.

4.7 SUMMARY

The physical place where an insect lives is its biotope. A species' habitat is the place where it lives relative to its behavioural ecology and life-history style. Biotopes can be changed but not destroyed. For a certain species of insect, its own habitat can be destroyed as the biotope is modified. For many insects, the pattern of the vegetation is an important determinant of their survival. This plantscape is a useful concept in terms of insect conservation biology.

At the smallest scale of measurement is the actual physical place where a particular life stage of an insect lives – its microsite. The microsite and its environmental conditions relative to the insect's behavioural ecology is its microhabitat.

These small-scale levels of site determination are intimately linked with larger scale processes. The biotope forms part of the landscape, an important level of measurement in insect and biodiversity conservation. There is a further hierarchy of scales from the landscape to the global level. All scales are interlinked.

At the global level, stresses are increasing, and these are likely to cause enormous insect genetic loss. The present rate of enhanced global warming is too fast for directional selection. Experimental evidence suggests that only the most temperature-tolerant of tropical species will survive a modest 2°C increase in overall temperature. These extinctions are likely to be compounded by multiple stresses on plants. A change in the global situation will change the plantscape, which will change the insect herbivores, their parasitoids and predators. Stenotopic and geographically restricted species will probably disappear, leaving the eurytopic and/or widespread species. Certain insect groups, such as the lycaenid butterflies, which are sensitive to biotope conditions and mutualistic relationships, are often restricted to small geographical areas, and will be especially affected.

Even at the level of ecosystems and landscapes, there will be compound changes in response to global climatic behaviour, but the exact details are difficult to determine at this stage. There is definite merit in starting monitoring programmes using insects as indicators of the changes. Key areas would be the tropics, especially along altitudinal gradients.

Nature reserves cannot move, although they may be able to expand. As primary production is tightly coupled to rainfall, and as most insects are closely associated with plants, there are likely to be profound changes in nature reserve insect communities over the next 100 years, as the plants and insects themselves are affected.

Different ecosystems and biotopes will respond in accordance with the severity of the change. Hyperacidic precipitation in the north-temperate region will have more localized effects, but most insects in this area may be relatively tolerant to climatic warming, being more adversity adapted than tropical species. There is likely to be a much more severe impact in the southern hemisphere with its large number of endemic species. But they too will not be affected to the same extent as the relatively climatically intolerant tropical species. Coastal areas will suffer an increase in sea level which will adversely affect coastal wetland fauna. Caves, often with unique stenotopic faunas, will be severely affected, as water patterns change and human population pressure increases. Mountains, often being wilderness or protected water catchment areas, may be the saving grace for some species that are able to move up the mountainside in response to warming. This assumes that their host plants are also able to do so, and that fragmentation of the landscape is not too severe so as to impede movement.

Other large-scale issues include insect migrations, in response to local and global effects which may change. Roosting sites that are protected phenomena, such as those of the monarch butterfly, will require careful monitoring and protection management.

—5

The fragmented landscape

The division between biological conservation and other
branches of conservation is no longer tenable. The
preservation of a prehistoric hill fort is inseparable
from that of the butterflies that inhabit that particular
kind of grassland on the ramparts.

Oliver Rackham

Lying in the open grave with tradition is biodiversity.
As crushed as were European wheat fields,
playgrounds for Allied and Axis tanks, we suffer, and
live with, the consequences of our actions.

Daniel H. Janzen

5.1 LANDSCAPE ECOLOGY

Mankind has geometrized the landscape. This has continued from the beginning of settled agriculture, and has generated the field of landscape ecology (Forman and Godron, 1986; Naveh and Lieberman, 1990; Turner and Gardner, 1991). The geometrical fragmentation is particularly evident in the agriculturally rich north-temperate lands (Wilcove, McLellan and Dobson, 1986).

Landscapes have three basic structures: the matrix, patches and corridors (Figure 5.1). Forman and Godron (1986) apply the following definitions:

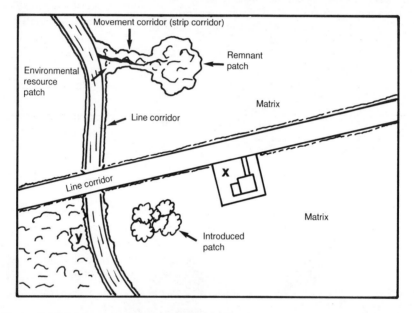

Figure 5.1 A landscape composed of three main elements: matrices, patches and corridors. On the right the farmhouse (x) is surrounded by an agriculturally disturbed matrix, which is bisected by a road line corridor. At the edge of the road is a line of vegetation, characterized by edge species. On the left, running from top to bottom, is a stream. The riparian vegetation along the stream forms a line corridor. In the bottom left-hand corner is a natural forest matrix, but with a disturbance patch area (y) where a few trees have been removed. Across the stream a patch of trees has been planted in the agricultural matrix. These trees constitute an introduced patch. On the other side of the road a remnant patch of native vegetation is joined to the woodland riparian vegetation by a wide strip corridor, which contains both edge and interior species, as does the patch to which it is connected. The patch of vegetation adjacent to the stream surrounds a small gulley which is rich in biota and detritus. This is an environmental resource patch. Some of the fauna moves back and forth from the stream edge along the strip corridor (making it a movement corridor) to the remnant patch.

Landscape. A heterogeneous land area composed of a cluster of interacting ecosystems that are repeated in similar form throughout. Landscapes vary in size, down to a few kilometres in diameter.

Matrix. The most extensive and most connected landscape element type present, which plays the dominant role in landscape functioning. Also, a landscape element surrounding a patch.

Patch. A non-linear surface area differing in appearance from its surroundings.

Corridor. A narrow strip of land that differs from the matrix on either side.

Points at which corridors coincide, such as the intersection of hedgerows, are called nodes. The degree to which landscape elements are connected to each other is the connectivity of these elements in the landscape [see Forman and Godron (1986) and Saunders and Hobbs (1991) for details]. The movement routes of animals between landscape elements can be mapped and made into continuity diagrams (Wood and Samways, 1991) (Figure 5.2).

For insect conservation biology, landscape ecology provides a useful conceptual framework for spatial and functional analysis relative to the habitats and behaviour of insects (Samways, 1989f).

5.2 MATRICES

As the matrix is the most extensive and most connected landscape element type present, it plays the dominant role in landscape functioning. It is also a landscape element often surrounding a patch of a different nature.

The most important, extensive and sensitive natural matrix in terms of insect biodiversity is the tropical rain forest. Its present rapid destruction is quite different from the gradual transformation over the centuries of the more established fragmentary nature of the European landscape.

Other major natural matrices are also important, because of their extensiveness. These include deserts, savanna, temperate conifer forests, taiga and tundra. North American forests have been removed at a high rate, but sustainable management plans are now being put into practice (Harris, 1984). Such sustainability through rotational cropping continually leaves refugia for biota. For insects, this principle has been applied to bugs (Morris, 1990a) and beetles (Morris and Rispin, 1987) of calcareous grasslands.

From a conservation biology viewpoint, the matrix may be highly degraded and disturbed. This is the case of the cartwheel effect of vegetation-bare ground caused by cattle congregating around water holes in savanna areas. Or the matrix may be a species-poor crop monoculture. This is the domain of the applied entomologist aiming to suppress the minority of species that thrive on well-tended host plants. The matrix may

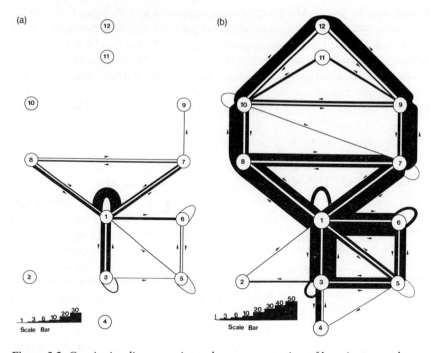

Figure 5.2. Continuity diagrams give a clear representation of how insects and other animals move across the landscape. These are two continuity diagrams of the southern African butterflies (a) *Acraea encedon* and *A. oncaea* (which cannot be distinguished on the wing), and (b) *Catopsilla forella*. The width of the lines and arrows refer to the number of butterflies flying in a particular direction between 12 distinct landscape features. The loops indicate the number of butterflies starting and ending their flight at particular landscape features: 1 = edge of pond, 2 = plane trees, 3 = cut grass, 4 = cut grass and flower beds, 5 = uncut grass, 6 = hollow in ground with long grass and forbs, 7 and 8 = open water of pond, 9 and 10 = forest edge, 11 = interior of forest 12 = above the forest. Most species, including these here, mostly used the edge of the pond, the forest edge and the hollow as flight paths. Most species moved in and out of the cut grass, reacting as if to a distinctly non-preferred biotope. (From Wood and Samways, 1991.)

even be insect-depauperate urbanization, surrounding a green-lung city park.

Matrices are vital determinants of insect diversity, ranging from the sterile urban environment (although there are almost always some ants and cockroaches) to the most species-diverse ecosystems on earth. The important point is to recognize that natural matrices, especially those of the tropics and southern hemisphere, which are high in endemics, are rapidly being eroded. For insect conservation biology, it is essential that more research be done in these extensive matrices, which are being removed, degraded or fragmented worldwide. In particular, there is an urgent need to catalogue the tropical endemics, and to identify areas of special interest within the megadiversity areas. K.S. Brown (1991) has begun to do so for butterflies of Brazil. There is an urgent need for more conservation biology orientated biogeographical studies such as those of Brown in other parts of the world. Already valuable background information exists, again mostly on butterflies, from elsewhere [e.g. Lepidoptera of the Indo-Australian region (Holloway, 1986)].

5.3 PATCHES

5.3.1 Disturbance patches

An island of disturbed land within an undisturbed matrix is a **disturbance patch**. Such patches happen naturally in the tropical forest when a large tree falls, taking with it epiphytes, creepers and smaller trees. This opens a patch of the forest on which certain insects thrive. The neotropical flightless grasshopper, *Microtylopteryx hebardi*, inhabits gaps in the rain forest (Braker, 1991). Survival of such a relatively immobile insect on such an ephemeral resource is possible because the gaps are dynamic. A new gap is often created at the edge of pre-existing gaps, making the appropriate successional stage almost always available at the small spatial scale. Other, often more mobile, insects invade and make use of the fallen and decaying trunks (Janzen, 1975).

Shure and Phillips' (1991) studies on natural disturbances in North American forests showed that arthropod abundance and community composition vary across the different sizes of forest openings. Arthropods from the surrounding forest readily occupy the smallest canopy openings (0.016 ha). All feeding guilds were well represented in these small openings and herbivore biomass and load (mg of herbivores per g of foliage) were much higher than in larger patches. In contrast, arthropod abundance and species richness were significantly lower in mid-size than smaller patches. Shure and Phillips (1991) suggest that the relatively sparse cover and high sunlight in mid-size openings may have promoted surface heat build-ups or soil surface/litter moisture deficits which restricted arthropod entry from the surrounding forest. Further, arthropod abundance and species richness were higher in large than mid-size patches (Figure 5.3). The greater vegetation cover in larger openings may have minimized the deleterious effects on arthropod populations.

Disturbance patches in a natural matrix, if left to regenerate, have a characteristic secondary plant succession that varies from area to area. In a study by Brown, Hendrix and Dingle (1987), which was in Old and New World grasslands, insect herbivores were major determinants of direction and rate of succession.

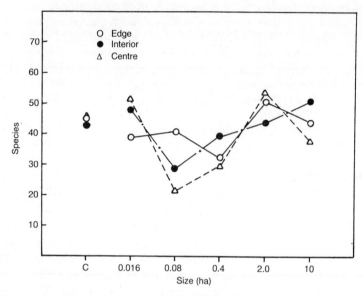

Figure 5.3 Arthropod species richness along the edge to centre gradient in North American forest openings of different sizes. (From Shure and Phillips, 1991.)

Most of the conspicuous disturbance patches are man induced. In traditional low-intensity shifting agricultural systems, there is little impact on the overall ecosystem matrix. But when the disturbance patch becomes extensive through wholesale forest removal, the patch becomes a degraded matrix with major impact on insect species' behaviour and survival.

Schowalter (1985) has pointed out that insect responses to changing conditions within patches contribute to individual fitness. They also appear to stabilize ecosystem productivity by (a) regulating plant–soil nutrient–light relations through changes in plant-age structure and/or species biomass relations, and (b) increasing reliability of disturbance and thereby mitigating its impact. This regulatory importance of insects appears to be a function of disturbance frequency and severity. Certainly, recovery by insect populations to severe natural disturbance such as storms and floods, can be quite rapid (Samways, 1989c). Also, where the disturbance is large and natural, as with volcanic eruptions, certain insects can be among the first colonizers (New and Thornton, 1988). They found that on Anak Krakatoa, Indonesia, a small nemobiine cricket, *Speonemobius* sp., is highly active on bare lava flows that reach 66°C during the daytime. This cricket is the dominant opportunistic feeder on the remains of wind-borne insects that land and are unable to survive on the vegetationless terrain.

5.3.2 Remnant patches

Remnants of pristine, near-pristine or ancient modified patches, whether heaths, moorlands, marshes or woodland, are extremely important insect reserves within a generally unfavourable agricultural and urban matrix. But there is a cautionary rider, in that natural or pristine landscapes hardly exist anywhere in temperate Europe, and this has an important bearing on the insect species present and their abundances (Peterken, 1991; J.A. Thomas, 1991). In Britain, many remnant sites are designated Sites of Special Scientific Interest, and may represent the last remnants of once widespread ecosystems (Moore, 1987a; Fry and Lonsdale, 1991). Such remains are often under enormous local pressure from their surroundings, such as building development, intensive farming and local pollutant emissions, as well as from the longer term threat of global atmospheric changes.

The ancient woodlands were formerly managed in a way that encouraged a wide variety of insects. Pollarding, for example, opened the tree canopy, while stands of undisturbed forest had trees of all ages and decaying logs, all promoting a suite of suitable habitats. Concurrent with the removal of many of the last remaining patches of forest have been changes in management of the woodlands over the past 50 years. This increasing insularization and biotope change has caused a massive decline in the ranges of insect species (Warren and Key, 1991). There is now an urgent need to modify management practices to encourage again the variety of insects that inhabit the various seral stages from early to late. Warren and Key (1991) also emphasize the value of open spaces and dead wood in forests, both for butterflies and for early successional species. Appropriate management is vital and urgent, whether for a coppice wood or for an old forest.

In the tropics, a similar fate is befalling the vast tapestry of biotope-sensitive, often endemic, insect life. Where the forest is being removed, the remnant patches need to be large so that interior forest conditions continue to prevail. C.D. Thomas (1991) found that in the wet lowlands of Costa Rica endemic butterflies are restricted to unmodified biotopes, with deforestation having considerable impact upon them.

Elsewhere in the world the picture is the same, with increasingly intense fragmentation of all types of landscapes taking place. In some areas the establishment of reserves, especially for game animals in Africa, means setting aside large remnant areas. These reserves, at least for small animals like insects, function more like a matrix than a smaller patch, being extensive and with a wide range of interior biotopes. Management of the plantscape is either natural from megaherbivore feeding and lightning-induced fires, or seminatural through fire management.

Figure 5.4 Patches of forest in an almost natural grassland matrix in Natal, South Africa. (a) Environmental resource patch of plant-rich, mixed woodland growing in the cool, moist ravines. (b) Introduced patch of *Eucalyptus* spp. surrounding and sheltering two cottages. The dots of individual trees between the arrows are *Pinus patula*, rogue seedlings from distant plantations.

5.3.3 Environmental resource patches

When a patch of vegetation is left behind because local conditions have allowed it to remain despite general environmental change, this is an environmental resource patch (Forman and Godron, 1986). This may be a mountain cleft, a localized bog or a desert oasis. With disturbance patches, the boundary between the patch and matrix is often sharp. But with environmental resource patches the boundary may be more gradual, and particularly rich in insect life. There are exceptions, and it is still a mystery why the remnant forest patches in the Drakensberg Mountains of southern Africa have such sharp boundaries, with plant and insect communities changing completely within a few metres (Figure 5.4). Interaction between natural grassland fire and topography seem to play a role.

The important point is that environmental resource patches are extremely important reserve areas for insects. Far more research is required to quantify this importance.

5.3.4 Introduced patches

Within a matrix of whatever sort, a patch of vegetation may be established.

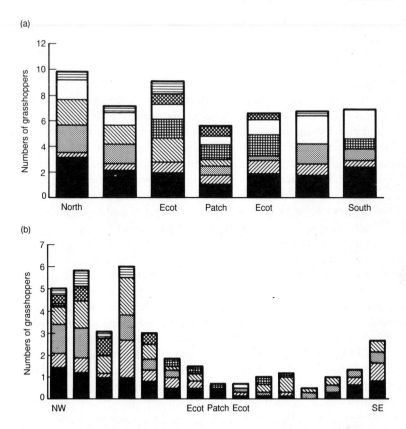

Figure 5.5 Mean grasshopper abundance (each type of shading represents a different species) per 5 m² in the ecotone (ecot) and up to 10 m either side of a small introduced patch of *Cupressus arizonica* trees in a grassland matrix. (a) Here the introduced patch had little impact on the overall grasshopper assemblage, in contrast to an introduced patch of *Pinus roxburghii* (b) which had a considerable depressant effect upon grasshopper abundance many metres either side of the patch, which here was 30 m rather than 10 m. (From Samways and Moore, 1991.)

This is an introduced patch. Relatively permanent disturbance patches, such as planting of exotic trees (Figure 5.4) can impact on the matrix beyond the edge. However, the influence depends on the type of tree. In South Africa, there is a strong, positive correlation between grasshopper and grass species richness. Exotic *Cupressus arizonica* tree patches increase orthopteran species richness and abundance, while exotic *Pinus roxburghii* and *P. elliotti* patches have an adverse impact up to 30 m into the grassland matrix (Samways and Moore, 1991) (Figure 5.5).

In many countries, plantations of exotic pines and eucalypts are replacing natural biotopes. In places, these plantations are directly risking the survival of localized threatened species, as with some threatened lycaenid butterflies in South Africa (Henning and Henning, 1989; Samways, 1993). The effect is sometimes direct through biotope replacement, or indirect by altering the local hydrological patterns.

Ormerod, Weatherley and Merrett (1990) have shown that larvae of the rare dragonfly *Cordulegaster boltoni* occur abundantly in Welsh moorland streams and in a deciduous woodland stream, but are never found in streams draining plantations of conifer forest. They found that biotopes of the stream margins were highly eroded and often absent from the conifer forest streams, providing generally unsuitable living conditions both in the water for the larvae and possibly for foraging by adults. Indeed, *Pinus* spp., although of great conservation value in their native areas (e.g. Fry and Lonsdale, 1991), can be highly suppressive for insects as plantations in foreign lands. Turin and den Boer (1988) showed this to be the case for carabid beetles, which are normally abundant in deciduous forests.

Some of the plantation companies are not insensitive to the need for protection of threatened biota, including insects. One company in South Africa has set aside land for the only known extant population of the lycaenid *Orachrysops ariadne*. It is uncertain what management procedures to undertake, although it does appear that the biotope is being overgrown as the grasses and forbs give way to increasing density of bushes. This is principally because the agricultural and forestry activities have reduced megaherbivore activity and possibly reduced the incidence of savanna burning.

The matrix–patch mutual influence depends not only on the type of introduced patch, but also on the species of insect involved and how it responds to the immediate microclimate. Ant assemblages, for example, are sensitive indicators of environmental conditions (Andersen, 1990), and their diversity changes sharply at patch edges (Figure 5.6).

5.3.5 SLOSS in terms of a landscape patch modelling

SLOSS is the acronym used to describe the debate as to whether the best strategy for species survival is to have a single large or several small remnant refuge patches as nature reserves (Diamond, 1975; Wilson and Willis, 1975; Wilcox and Murphy, 1985; Spellerberg, 1991). The debate has generated considerable controversy, and among other approaches has been formulated in terms of maximizing species richness or in terms of minimizing extinction rates (Burkey, 1989). The crucial conservation biology question hinges on whether the function of a nature reserve should be to support more species, or whether it is more important to weight species so that the reserve contains more species that would become extinct in the absence of the reserve.

Burkey's (1989) stochastic simulation model points to time as being crucial

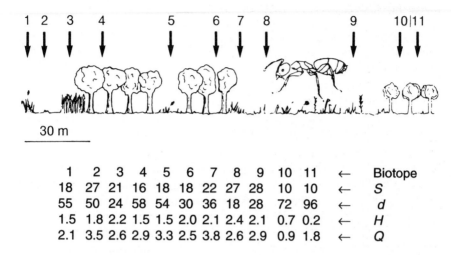

1	2	3	4	5	6	7	8	9	10	11	←	Biotope
18	27	21	16	18	18	22	27	28	10	10	←	S
55	50	24	58	54	30	36	18	28	72	96	←	d
1.5	1.8	2.2	1.5	1.5	2.0	2.1	2.4	2.1	0.7	0.2	←	H
2.1	3.5	2.6	2.9	3.3	2.5	3.8	2.6	2.9	0.9	1.8	←	Q

Figure 5.6 Diversity of epigaeic ant species associated with 11 disturbance biotopes (= habitats when species' behaviours are also considered) in lowveld South Africa: 1, mixed grass and forbs; 2, bare soil of farm track; 3, rank grass; 4, full canopy in mature citrus orchard; 5, clearing in mature citrus orchard; 6, partial canopy in mature citrus orchard; 7, stand of rank grass; 8, almost undisturbed, thin short grass; 9, lightly disturbed grass of various heights; 10, partial canopy of young citrus orchard; 11, full canopy of young citrus orchard. S = species richness; d = Berger–Parker dominance index, emphasizing the dominant species; H = Shannon–Wiener index, emphasizing the various commonest species; Q = Interquartile index, emphasizing the mid-range species on the rank-abundance curve. The highly disturbed biotopes (young citrus orchard, nos. 10 and 11) were species poor and strongly dominated by one species, while the least disturbed, near-natural, open grassland biotopes (particularly 8 and 9) were species rich and highly equitable, with sharp changes occurring at the biome boundaries (From Samways, 1983a, 1989c.)

to how the risk of extinction changes with fragmentation. For short to moderate timespans, the probability of extinction increases exponentially with degree of fragmentation. When the timespan is long relative to the size of the reserve, the extinction rate becomes sigmoidal with increasing timespan or decreasing reserve size. The curve becomes steeper, with eventually only the sheer incline leading to the truncated sigmoidal curve remaining. Burkey's (1989) models further show consistently that a species is more likely to survive in a continuous tract of natural habitat than in one that is subdivided into isolated patches, again emphasizing the value of extensive natural matrices. Migration and also less directed movements greatly reduce the probability of extinction in fragmented landscapes.

Even small reserves are home to many insects, providing there is shelter for all the life stages: eggs, larva and adult. But the smaller the reserve, the increasingly severe is the impact at its edge during adverse external conditions.

Another factor in the SLOSS argument is that large reserves may have minimal extinctions compared with small reserves. Chance environmental impacts are less likely to cause extinctions in large reserves, yet many small reserves, in contrast, may spread the risk.

The SLOSS debate, which originally arose out of the theory of island biogeography, has provided a useful conceptual framework. But it has been little tested empirically and probably is too generalized to cover the requirements of all species. Spellerberg (1991) points out that large reserves are not always the most appropriate. Of most importance is to conserve as much as possible, as long as possible. This has developed into the 'SLOPP' ('single large or plentiful patchy') debate which argues that it may be better to divide an area into very many small patches, each with many individuals, rather than a single large area. The idea is that the number of equal-sized biotope fragments should be equal to the square root of the carrying capacity for the target species. However, this still leaves the problem of different species having different requirements which do not obey a general statement. Also, this contrasts with Burkey's (1989) models, again urging experimental evidence one way or the other.

5.4 CORRIDORS

5.4.1 Line corridors

These include most paths, roadsides, thin hedgerows, property boundaries, drainage ditches and irrigation channels, all of which are narrow bands essentially dominated throughout by edge species (section 5.6).

The hedgerows of Europe are well known to have a community structure commensurate with their age (Pollard, Hooper and Moore, 1974). They are, or certainly have been, an important and major feature of the fragmented agricultural landscape (Zanaboni and Lorenzoni, 1989). Butterflies such as the gatekeeper (*Pyronia tithonus*) are probably much more common now than before the Neolithic forest clearances, as there are more sunlit edges, both around woodland fragments and along hedges, providing a suitable biotope.

Hedges provide refuge and food for insects (Lewis, 1969a), especially during adverse conditions. They also have an influence on movement patterns of some insects in association with wind patterns (Lewis, 1969b). Management patterns also have a strong influence on insect presence or absence and distribution patterns. In recent years, there has not been sufficient replacement of filled or dying trees. In addition, there has been stunting of naturally regenerating saplings by annual cutting with mechanized flails. The loss of insect species associated with oak trees has been particularly deleterious to insect conservation because of the large number of species dependent on these trees (334 species of insects and mites feeding on *Quercus petraea/robur*)(Fry and Lonsdale, 1991).

Hedges in the agricultural landscape can be relatively good reservoirs of forest-dwelling species, as shown by Burel and Baudry's (1990) studies on carabid beetles in northern France. When the forest carabid species are viewed individually, they can be divided into three groups according to distance from the forest: (1) forest core species occurring only 100 m along the hedgerow from the forest, (2) forest peninsula species which penetrate the hedgerow up to 500 m from the forest and (3) forest corridor species, which use hedgerows as corridors and are found at any distance from the forest (15 km). However, Burel and Baudry (1990) caution against extrapolating from one taxon to another. In their case, plants and carabids gave quite different landscape distribution patterns. Such discrepancies are also likely for different insect taxa. These differences often may be at the species level rather than at a higher level. A certain beetle, let's say, may behave in a similar way as a particular bug. But not all beetles will be different *en bloc* from all bugs.

Figure 5.7 Species-poor, drainage dykes in an intensive agricultural matrix in Zeeland, The Netherlands. Can lack of insect conservation concern be criticized here? It is land reclaimed from the sea where these insects would not have occurred naturally. The Odonata, at least, are all widespread, eurytopic common species.

Human pathways can be beneficial or harmful depending on type and usage. The bare ground of a minimally used path is often a sunning or oviposition site. The plant edge species may also benefit specific insects. For others they may be an ideal foraging area, particularly for ants. Alternatively, paths are a physical barrier exposing the insect to unaccustomed harsh conditions and predators.

Drainage channels and ditches are also often important reservoirs for

(a)

(b)

species, especially eurytopic zygopterans. However, the advantages are often offset by the fact that they are draining wetlands, or may receive nutrient-rich run-off from agricultural land (Figure 5.7).

Fences do not have the same devastating effect on insects as on large vertebrates. In fact, metal and wooden fence poles can be home to some insects such as *Crematogaster* spp. ants.

Roads are line corridors that can cause high mortality where traffic volume is high. In 1989, 100 m of Tennessee roadside was a graveyard for over 120 traffic-killed butterflies. This number included only those that were countable and excluded those that had already rotted or had been carried away by ants, as well as all the other insect groups.

Roads are generally whole or partial barriers, depending on the intensity of traffic and the mobility of the insect (Figure 5.8).

In southern France, populations of the extremely rare bush cricket, *Platycleis fedtschenkoi* subsp. *azami*, have been fragmented by roads, yet the airport grounds in turn have provided refuge areas (Samways, 1989a). This species is flightless, yet other highly localized species e.g. *Platycleis falx*, readily fly across the smaller roads. In short, the barrier is relative to the mobility and biotope preference of the insect, varying from a complete barrier to a preferred biotope (Tables 5.1 and 5.2).

Mader (1984) has clearly shown that tarred roads can be highly influential on carabid beetle movement patterns. For stenotopic woodland species, roads are virtually a complete barrier. For eurytopic species this is less so, for low-vegetation species less so again. Nevertheless the roads are still highly inhibitory to movement across them. Even untarred forest roads are highly inhibitory.

Railway cuttings have always been known as insect-rich areas in temperate lands, especially the sunny south-facing slopes in the northern hemisphere. Today, motorway verges and banks are also providing extensive new areas for habitation. The difference is the frequency of the transport, with motorways being treacherous to cross for all but the highest fliers. There is also a move to ecologically landscape these verges, which are often extensive

---

Figure 5.8 The fragmented landscape around Montpellier, southern France. (a) This fragment is one of the last refuges of the flightless bush cricket, *Platycleis fedtschenkoi* subsp. *azami* which cannot cross the tarred road with heavy traffic (arrowed). The road is a corridor, and a barrier, both by its nature and relative to the mobility of the insect (see Tables 5.1 and 5.2). This contrasts with another site (b) where a track (arrowed) that is infrequently used by vehicles presents no barrier to the flightless immatures or flying adults of *Platycleis affinis* and *P. intermedia*. The strip of grass at the base of the wall is particularly attractive to immatures. Young *P. affinis* adults then spread to the barley field to the left in this photograph, and the *P. intermedia* adults move to the bushes adjacent to the wall.

Table 5.1 The landscape elements present in the vicinity of Montpellier, southern France. This list is not necessarily exhaustive, and the maximum patch size is relative to the maximum home or flight range of the subject animals, which were here dectine bush crickets. (From Samways, 1989a)

Homogeneous line and water corridors	Matrices (surrounding patches <150 m diameter)	Patches (<150 m diameter)
Untarred roads	Etangs	'Sansouire'
Tarred roads	'Sansouire'	Trees and bushes
Roadside verges	Mixed grasses and forbs	Grasses and forbs
Walls	'Garrigue'	Long, rank grass
Drainage ditches	Vineyards	Vineyards
Streams and rivers	Wheatfields	Wheatfields
Boundaries	Buildings	Rock outcrops
		Buildings

enough to establish a varied plantscape suitable for a wide range of insect species (Fry and Lonsdale, 1991) (Figure 5.9).

For the insect conservation biologist, line corridors are often extremely important linear patches relative to the insects' home ranges and in directing movement. Line corridors are principally not of prime genetic conservation wealth, as the insects present are mostly the more widespread, physiologically adaptable species, not the taxonomically isolated or stenotopic rarities. However, the insect biomass of such corridors is encouragement to their vertebrate predators, which in turn may be of conservation value.

5.4.2 Strip corridors

A strip corridor has an edge effect on either side, but the strip is wide enough to contain an interior environment (Forman and Godron, 1986). These are wider than line corridors but intergrade. In fact, very wide ancient hedges and wide powerline cuttings and super highways may be considered in this category, with differential positive and negative effects depending upon habitat circumstances. Forest rides (wide bridle paths) of ancient origin are important for some insects (Warren and Key, 1991).

In general, strip corridors are of greater insect conservation interest than line corridors, simply because of the ecotonal variation towards interior plant species and environmental conditions. Whereas line corridors may be seen as long, thin patches, strip corridors are more like a larger patch with an interior.

5.4.3 Linkages

Harris and Gallagher (1989) and Harris and Scheck (1991) have made the

Table 5.2 The flight range and biotope of bush crickets (Tettigoniidae), of the subfamily Decticinae, occurring in the vicinity of Montpellier, southern France. (From Samways, 1989a)

Species	Approx maximum flight range (m)	Biotope of adults	Local distribution
Decticus albifrons (Fabricus)	150	Dry, mixed long grasses, low bushes and forbs	Widespread in the coastal hinterland around Montpellier
Decticus verrucivorus (Linnaeus) ssp. *monspeliensis* (Rambur)	0	Rank, long grass beside farm tracks	Apparently extinct; formerly one locality east of Montpellier
Platycleis affinis Fieber	<50	Dry, mixed grasses and forbs	Hinterland around Montpellier
Platycleis albopunctata (Goeze)	<25	'Garrigue' chapparal	The 'Garrigue' north of Montpellier
Platycleis fedtschenkoi (Saussure) ssp. *azami* (Finot)	0	Rank, long grass and forbs	Highly localized patches immediately south of Montpellier
Platycleis falx Fabricius	<30	Dry, mixed long grasses, low bushes and forbs	Localized patches on the coastal hinterland
Platycleis intermedia (Serville)	<100	Dry, mixed long grasses and forbs under or adjacent to bushes	Hinterland around Montpellier; not coastal nor montane
Platycleis sabulosa Azam	<100	*Salsola* and *Tamarix* 'sansouire'	Around the coastal lagoons south and south-east of Montpellier
Platycleis tessellata (Charpentier)	<5	Dry, mixed short and long grasses and forbs	Coastal hinterland around Montpellier

Note: P.f. azami was formerly known as *Metrioptera azami* Finot.

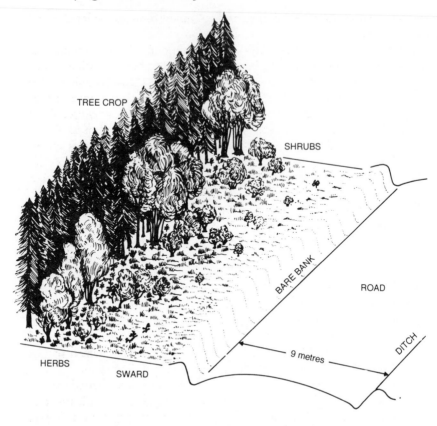

TREE CROP

SHRUBS

BARE BANK

ROAD

DITCH

9 metres

HERBS

SWARD

Figure 5.9 Ecological landscaping of a roadside verge with a varied plantscape suitable for a suite of insect species in Britain (From Fry and Lonsdale, 1991.)

strong point that corridor linkages between reserve areas are important as movement corridors for animals. Movement may be necessary for various reasons, from seeking a mate to finding new pastures. At the genetic level, movement corridors permit natural gene flow without having to resort to artificial translocation of individuals. Such movement corridors may be broken, but the breakpoints must be small enough for the species concerned to cross. The tolerance gap will vary from one species to the next, and for the different age groups, and possibly for the different sexes.

Thomas (1983a) found that the adults of the Adonis blue butterfly (*Lysandra bellargus)* readily flew 250 m over open calcareous grassland, but appeared not to do so over 100-m gaps of agriculturally improved grass, hedge and scrub. Similarly, Warren (1987a,b) found that the heath fritillary butterfly (*Mellicta athalia*) moved along woodland rides linking clearings (their breeding sites) within continuous blocks of woodland, but almost never crossed even short stretches of farmland to move from one isolated wood patch to another.

Landscape element pattern plays an important role in determining the flight responses of certain insects. In a study of various butterfly flight paths, Wood and Samways (1991) showed that in a South African botanic garden there were certain preferred flight paths for most species. Ecotones such as the edge of an area of water and of a forest were the most heavily used flight corridors. A disturbed and naturally regenerated mixed vegetation patch was also an important flight pathway. A stand of large exotic plane trees and an expanse of cut grass were highly negatively influential, causing the butterflies to change direction. There were differences between species, with some but not others finding the water stand a major barrier to cross.

Movement corridors may not be entirely beneficial. For some insects and other animals such linkages may be a channel of prosperity for certain predators and parasitoids which may effectively block movement of the host. The linkages may also allow the spread of pathogens from one patch to another. The significance of these trade-offs is a rich field of investigation for insect conservationists. These factors interplay with corridor width, length, shape (funnel-shaped versus straight) and degree of departure from linearity (Soulé and Gilpin, 1991). Each species with its characteristic physiology and behaviour under different seasons and environmental conditions will respond differently. Conservation objectives would need to consider which species and communities should be the particular conservation targets.

Keals and Majer (1991) studied the ant assemblage along a road conservation corridor in Western Australia. The verges supported a wide range of the ant species that occurred in the region, but species richness was low in the absence of native vegetation. Even where native vegetation was present, verges needed to be wide [>20 m, as opposed to <5m (narrow)] for the microclimate to be appropriate to support the richer fauna.

5.4.4 Corridors, patches and matrices relative to biotope preference

Table 5.1 lists the landscape elements present in the Montpellier area of southern France. The maximum patch size is determined here by the minimum flight range (home range) of the most mobile of the subject animals (decticine bush crickets) (Table 5.2). As with corridors (Table 5.3), the relative significance of the patches (Table 5.4) and matrices (Table 5.5) depends on the biotopes that they contain relative to the lifestyle of the species in question. Among these closely related species there is a vast range of preference, from the landscape element providing no or little refuge, to the other extreme in being a preferred biotope. When such a phylogenetically close assemblage as these tettigoniids shows such a wide range, it emphasizes that extrapolation to the community level involves the conservation of a

Table 5.3 The line corridors and their impact relative to the mobility and biotope (see Table 5.1) of the nine decticine bush crickets occuring around Montpellier, southern France (From Samways, 1989a)

	Untarred roads	Tarred roads	Roadside verges	Drainage ditches	Streams and rivers
D. albifrons	4	4	6	6	3
D. verrucivorus ssp. monspeliensis	2	2	8	5	1
P. affinis	4	4	7	5	3
P. albopunctata	3	3	4	5	2
P. fedtschenkoi ssp. azami	2	1	5	8	1
P. falx	4	4	6	6	3
P. intermedia	4	4	7	5	3
P. sabulosa	4	4	5	5	3
P. tessellata	2	2	6	5	1

1 = Almost complete barrier, without assistance; 2 = partial barrier; 3 = minor barrier; 4 = no or very minor barrier; 5 = marginal biotope; 6 = refuge; 7 = preferred biotope; 8 = the only suitable biotope.

diverse range of microsites, biotopes and plantscapes for survival of all the species and their interactions. Both for these bush crickets (Samways, 1976, 1977b) and for British butterflies (J.A. Thomas, 1991) these landscape elements must accommodate conditions for all life stages of the species, to ensure the species' continued survival in the area.

5.5 EDGES AND ECOTONES

With increasing fragmentation of the landscape, edges of patches of land are becoming proportionately greater relative to interiors.

Three, not completely mutually exclusive, terms are useful relative to landscape patterns and modification.

Edge. This is the physical boundary between two plant communities, or plantscapes, or between a plantscape and a disturbance area such as a ploughed field or a road. It implies a sharp boundary caused by human modification to the landscape.

Edge effect. This is the **influence** of the edge. It implies a **change** in community structure and in relative abundance of species when sampling from the edge to the 'interior' of the matrix on one side, and to the interior of the patch on the other. Traditionally, the edge effect meant an increase in numbers of species because of the intergrading of two biotopes (Leopold, 1933). It is preferable, however, to limit the definition simply to the **change** in species at the edge (see Lovejoy *et al.*, 1986).

Table 5.4 Relative significance of patches in supporting populations of the nine species of decticine bush crickets occurring around Montpellier, southern France (From Samways, 1989a)

	'Sansouire'	Trees and bushes	Grasses and forbs	Long rank grass	Vineyards	Wheatfields	Rock outcrops	Buildings
D. albifrons	3	1	6	3	3	3	2	2
D. verrucivorus ssp. monspeliensis	2	1	3	6	2	2	1	1
P. affinis	3	1	6	3	3	3	2	2
P. albopunctata	2	6	3	3	2	2	2	2
P. fedtschenkoi ssp. azami	2	1	4	6	2	2	1	1
P. falx	3	2	6	5	2	2	2	1
P. intermedia	3	6	4	3	3	2	2	2
P. sabulosa	6	3	3	2	2	2	1	2
P. tessellata	3	2	6	3	3	2	2	2

1 = None or very little refuge; 2 = limited refuge; 3 = limited penetration into patch; 4= partial geographical refuge; 5 = extensive refuge; 6 = preferred biotope.

Table 5.5 Relative significance of matrices in supporting populations of the nine species of decticine bush crickets occurring around Montpellier, southern France (From Samways, 1989a)

	Etangs	'Sansouire'	Mixed grasses and forbs	'Garrigue'	Vine-yards	Wheat-fields	Build-ings
D. albifrons	1	3	6	5	3	3	3
D. verrucivorus ssp. monspeliensis	1	2	3	2	2	2	1
P. affinis	1	3	6	4	3	3	4
P. albopunctata	1	2	2	6	2	2	2
P. fedtschenkoi ssp. azami	1	2	3	1	2	2	2
P. falx	1	3	6	5	2	2	2
P. intermedia	1	3	2	5	3	2	4
P. sabulosa	1	6	3	2	2	2	3
P. tessellata	1	3	3	4	3	2	2

1 = None or very little refuge; 2 = limited refuge at boundaries; 3 = limited penetration into matrix; 4 = partial geographical refuge; 5 = extensive refuge; 6 = preferred biotope.

Ecotone. This is the narrow overlap zone between adjacent plant communities. It can refer to a natural gradation of communities, or to human-created boundaries. Ecotone refers to the changes in both the plantscape and the fauna. It differs from **edge effect** in making no assumption about the abruptness of change in the plantscape.

These factors are highly significant for small animals such as insects as there are abrupt changes in light regimen, substrate, water conditions, species composition and other factors that generally characterize the edge (Wilcove, McLellan and Dobson 1986; Jägomagi *et al.*, 1988). As well as affecting living conditions both directly and indirectly through influence on food plants (MacGarvin, Lawton and Heads, 1986; M.S. Warren, 1987a,b), flight paths are also modified (Wood and Samways, 1991).

The edge effect and ecotone are not clearly definable discrete units. Each animal species responds differently to the boundary and the environmental characteristics of the spatially variable conditions of the plantscape. In addition, biotic interactions also influence the species distribution patterns. These include parasitism, mutualism, intraspecific behaviour and interspecific competition. Populations of species shift as these abiotic and biotic variables change at the boundary between plantscapes.

Ecotones occur naturally on a small scale from tree fall, on a larger scale from rock falls and on an even larger scale in the tension zone between fire-prone savanna and the remnant forest patches in montane clefts (Figure 5.4a).

With geometrization of the landscape, ecotones have become much more widespread. Also, they often run parallel either side of a corridor between large disturbance patches, as, for example, grass firebreaks between blocks of planted forest. Such corridors are intimately related not only to the animal's movement pattern, but also to the extent of the ecotone relative to the corridor interior.

In general wildlife terms, Soulé and Gilpin (1991) have developed a preliminary model of movement corridor simulation and have proposed that the optimal width of a corridor is a function of (a) the ratio of capability of edge versus interior and (b) the average distance moved by the animal within a time limit. The 'optimum' corridor width depends on the strength of the edge effect, i.e. the higher mortality rate resulting from too much edge reduces the value of the corridor. This means that species that are averse to the edge in favour of the interior, must move along the interior as fast as possible to reach the next patch. A wide, short, straight corridor would be the most favoured. For edge species more or less the converse applies. For ecotonal insects, other factors also operate, such as geographical aspect and availability of nectar sources.

Fry and Lonsdale (1991) provide many useful management guidelines for insect conservation at woodland edges in Britain. Much more research is required on edge effects in tropical and subtropical countries. The long history of fragmentation in temperate countries has allowed maturity and blurring of the divisions between landscape elements, making gradual ecotones commonplace. Harris (1988) and Yahner (1988) suggest investigating the beneficial and harmful effects of edges on wildlife management decisions. This applies to many taxa and communities, including insects, which are so readily influenced by this plantscape fragmentation.

Some initial model simulations have revealed that, for any edge-sensitive species and biotope type, there exists a critical range of fragment sizes in which the impacts of edge effects increase almost exponentially. To obtain this critical size range, it is necessary to have an empirical measurement of the 'edge function', i.e. response of taxa to induced edges (Laurance and Yensen, 1991). Apart from the work of Samways and Moore (1991), who found the edge effect for grasshoppers to be as much as 30 m, there is little information on how insect communities change with the edge effect. It can be influential for plants. Laurance (1991) found that for Australian tropical forest remnant patches, forest species disturbance was evident up to 500 m inside the patch, with striking disturbances up to 200 m from the edge. This suggested to Laurance that isolated nature reserves in NE Queensland must exceed 2000–4000 ha, depending on reserve shape, to ensure that up to 50% of the reserve remains unaffected by the fragmentation activities. How do the insects track these changes?

Certain insects, with characteristically small home ranges and particular biotope requirements, are well suited to testing such models. Also, the models have value in providing practical guidelines. Laurance and Yensen (1991) point out that confusion between area- and boundary-related extinctions arises because edge and area effects both become increasingly important in small fragments.

In addition to area, fragment shapes are an important determinant of edge effects. So Laurance and Yensen (1991) suggest researchers use at least three characteristics when attempting to explain abundances of species in fragments:

1. Interior species that are sensitive to edges should be most strongly and positively correlated with core areas;
2. Other taxa that depend on primary biotope but are not sensitive to edges should be positively correlated with total areas;
3. Edge species, especially those with small home ranges, should be positively correlated with total length of fragment edge.

Other factors must also be considered when describing the community structure of such patches. These include the isolation and spatial arrangement of fragment patches, as well as permeability of edges and location of dispersal sinks or source pools (Rapoport, 1982; Stamps, Buechner and Krishnan,1987; Laurance and Yensen, 1991).

5.6 TOPOSCAPE

The topographic landscape, or toposcape, has important conservation biology implications, especially for ectotherms such as insects. The shape of the land surface provides temporal and spatial perspectives for examining soil, vegetation and aquatic ecosystems, and for interpreting ecosystem processes (Swanson *et al.*, 1988). There are four classes of topographic landform effects on ecosystem patterns and processes, illustrated in Figure 5.10.

The landforms of the toposcape of the Natal Drakensberg Mountains in South Africa have been shown to have major conservation significance for

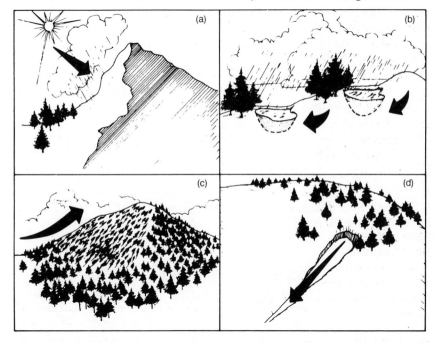

Figure 5.10 Examples of the four classes of landform influence on ecosystem patterns and processes. (a) Class 1. Topographic influences on rain and radiation (arrow) and their shadows. (b) Class 2. Topographic control of water input to lakes. Lakes high in the drainage system may receive a higher proportion of water input by direct precipitation than lakes lower in the landscape where groundwater input (arrow) predominates. (c) Class 3. Landform-constrained disturbance by wind (arrow) may be more common in upper-slope locations. (d) Class 4. The axes of steep concave landforms are most susceptible to disturbance by small-scale landslides (arrow). (From Swanson *et al.*, 1988.)

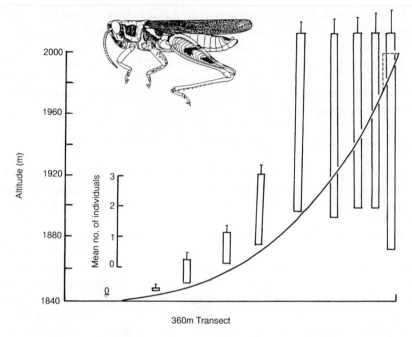

Figure 5.11 Mean number of grasshoppers in 5 m² quadrats on the eastern side of a grassy slope rising up to a craggy peak (a krantz) which provideds warmth and shelter for plants and insects in the Drakensberg Mountains in Natal, southern Africa. This emphasizes the significance of topography in insect behavioural ecology and for making decisions on conserving areas that take the vertical dimension of the landscape into account (From Samways, 1990b. Reprinted with permission of the Society for Conservation Biology and Blackwell Scientific Publications.)

montane grasshoppers (Samways, 1990b). Hilltops act as thermal refugia from winter cold air drainage. Additionally, the increased insolation on the eastern and northern sides of these southern hemisphere hilltops, compared with the western and southern sides, attracts the grasshoppers. Crevices in the hill summits provide further microrefugia, both for plants as well as for the insects (Figure 5.11).

Similarly, Murphy and Weiss (1988) used a topographic map of the biotope for long-term monitoring of the bay checkerspot butterfly, *Euphydryas editha bayensis*. This threatened butterfly makes use of the toposcape for survival, with the distribution of larvae changing between years, shifting from cool slopes to warmer slopes as the population size increases. These shifts affect the developmental phenology and timing of adult emergence.

Topography is additionally important for many insects to carry out behavioural activities. Some butterflies engage in 'hilltopping', principally as a mate-meeting mechanism (Alcock, 1987). Dragonflies, particularly some of the powerful fliers, take up feeding beats along ridges and small hills for capturing prey. Some other species of dragonfly make use of the thermal characteristics of the sheltered and rocky landscape to gain warmth for prolonged foraging (McGeoch and Samways, 1991).

5.7 SUMMARY

Man's increasing geometrization of the landscape through agriculture and urbanization has evoked the field of landscape ecology. The landscape, which is a heterogeneous land area made up of a cluster of ecosystems, has three basic structures: the matrix, patches and corridors.

The matrix is the most extensive and most connected landscape element, playing a dominant role in landscape functioning. Natural and minimally disturbed matrices are important for all biodiversity, from the species-rich tropical forests to the deserts, in view of their extensiveness. But most matrices, particularly in temperate countries, are now fragmented. The fragmentation, when minor in extent within a matrix of undisturbed land, results in disturbance patches, which are often naturally formed and are important for successional insect species. Remnant patches are the converse, being islands in a disturbed matrix. These are important refugia for many insects. Environmental resource patches are patches of vegetation left behind despite environmental change. A mountain cleft is an example, and such patches are generally important for both plant and insect conservation. Introduced patches are permanent disturbance patches, such as an exotic tree plantation, and are generally harmful to the local insect community.

The single large or several small (SLOSS) reserves debate is a conceptual component of landscape ecology. There has been considerable discussion on the merits and demerits of SLOSS for maximizing species richness or for minimizing extinction rates. The crucial insect conservation question is whether a reserve should support more species or weight species so that the reserve contains more species that would become extinct in the absence of the reserve. Extinction increases with the degree of fragmentation, but migration greatly reduces the chances of extinction, emphasizing the importance of movement corridors between suitable patches.

Line corridors are mostly dominated throughout by edge species. Most paths, roadsides, ditches, etc., are line corridors. They often provide refuge for many insect species, although not generally for interior species. Fences, although limiting the movement of large vertebrates, have little impact upon insects, although roads can be devastating barriers to insect fauna. Strip corridors are wider than line corridors, and have an interior environment, making them rich in insect species, both edge and interior ones. Corridors are often linkages, permitting movement from one patch to another. Such linkages play an important role in lessening the impact of fragmentation, but may have some adverse effects, such as being refugia for predators and parasitoids, and even possibly a highway for pathogens. Continuity diagrams are useful in illustrating the movement pathways of insects.

Edges and ecotones are highly significant for insects whose life-history style is shaped by the habitat. Woodland edge management is well known in insect conservation circles in Britain. This contrasts with the knowledge in tropical areas where fragmentation of the forest can result in an impact at least upon plant species up to 500 m into the patch. Far more research is required on how insect communities change at natural ecotones and how they respond to edges as the landscape is fragmented.

The toposcape or topographic landscape is important in insect conservation biology in providing refugia on high ground against cold-air drainage, and in permitting changes in behaviour pattern relative to population level variations.

6

The disturbed landscape

Up on the downs the red-eyed kestrels hover,
Eyeing the grass.
The field-mouse flits like a shadow into cover
as their shadows pass.
Men are burning the gorse on the down's shoulder;
A drift of smoke
Glitters with fire and hangs, and the skies smoulder,
And the lungs choke.

John Masefield

God will not ask thee thy race
Nor thy birth
Alone he will ask of thee
What hast thou done with the land I gave thee?

Persian Proverb

6.1 DISTURBANCES

Disturbances may be small and localized over a few metres, or widespread, natural or man-induced disruptions changing the appearance of the landscape. White and Pickett (1985) define a disturbance as a relatively discrete event in time that disrupts ecosystem, community or population structure and changes resources, substrate availability or the physical environment. Such disruptions may vary in distribution, frequency, predictability and severity. They may also be synergistic with other disturbances, such as long periods of drought, or may increase the risk of a fire outbreak.

As Pimm (1986) has pointed out, most of the earth is disturbed and dominated neither by pristine ecological communities nor, on the other hand, by species on the brink of extinction. The ecological communities are fragmented, harvested and polluted owing to the disturbance stresses caused by man. This has become increasingly so with the global atmospheric impact overlying the on-the-ground changes.

Communities and ecosystems are successional, gradually changing over time. These systems are subject to so many abiotic and biotic influences, and these are often so complex, that the familiar term 'stability' has little meaning in real conservation terms. Disturbances abruptly interrupt the successional gradualism, with the simpler systems tending to be more resilient. Resilience here refers to how quickly the system returns to its former state. Additionally, some systems may be more resistant to change, with species proportionate abundances readily returning to their 'equilibrium' proportions.

In insect conservation biology, little is known of how the relative abundances of insects change with disturbances, except through some detailed autecological and pollution studies, particularly in Britain. How many species are becoming extinct because of these disturbances? We readily see the plant community changing, and so presumably the insect community must also change. How many species become lost from the system? When does the time come when a key species is lost causing a cascading effect on other species? Empirical evidence is almost entirely lacking, although observations suggest that a suite of pressures, including invasive plants and insects, and also unplanned fires, can be damaging to the native insect community (Giliomee, 1992). There are also glimpses of a converse situation in which exotic insect populations soar to proportions where they overexploit their resources, yet this does not occur in their native home where they are just another moderately abundant species in the whole species complex. Red scale (*Aonidiella aurantii*) is such an insect which is not generally as common on its native *Citrus* spp. host in China as it can be elsewhere. In South Africa however, it is prevented from overexploiting its resources by parasitoids, *Aphytis* spp., at lower densities and by a key predator, the ladybird *Chilocorus nigritus*, when it escapes from parasitoid suppression (Figure 6.1). Various aspects of

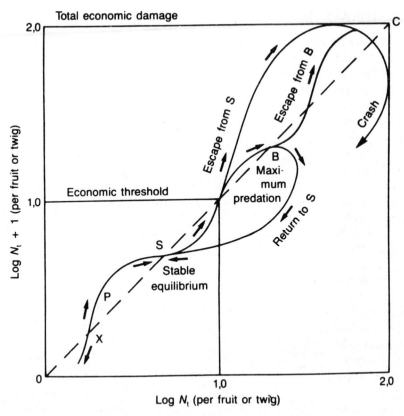

Figure 6.1 Population growth rate curve for red scale *Aonidiella aurantii*.
Normally the scale insect increases (P) to be maintained at a low stable equilibrium
(S) by the parasitoids *Aphytis* spp. In economic terms, the red scale population is
maintained below the economic threshold. When disturbance occurs, e.g. when
inappropriate pesticides are used against other pests, the population growth rate of
this key insect is such that it escapes the regulatory influence of parasitoids but is
then exposed to predation by the ladybird *Chilocorus nigritus* (B), which rapidly
returns red scale to S. The ladybird seeks other patches, while the parasitoids
re-exert regulatory influence. If an insecticide such as an inappropriate insect
growth regulator or organochlorine is used, or if environmental conditions are too
adverse (i.e. too dry), the red scale may escape from B and totally overexploit
resources, crashing to low levels as resources are lost. Only rarely are conditions
such that there is a total crash to extinction (X), although this can happen in
localized patches. (From Samways, 1988a.)

the morphology, physiology, phenology and behaviour of these important
species, as well as other species and weather conditions, influence the
proportionate population levels. The point is that red scale rises in abundance

status only when there is a severe disturbance (e.g. use of a persistent insecticide) to the system.

6.2 EXTINCTION VORTICES ASSOCIATED WITH MINIMUM VIABLE POPULATIONS

6.2.1 Extinction risks

Gilpin and Soulé (1986) provide a conceptual framework in which population vulnerability analysis (PVA) estimates the minimum viable population (MVA) before heading for extinction. With fragmentation of the landscape, such a framework provides a cornerstone for appreciating the problems facing insects. Gilpin and Soulé (1986) identify three interacting fields of analysis in population biology relative to PVA: population phenotype, environment and population structure, and fitness.

Extinctions themselves can be classified into deterministic and stochastic types. Deterministic extinctions result from an ongoing force from which there is no escape. An example would be the impact of global warming on small lycaenid butterfly populations living on mountain peaks. Destruction of a cave containing an endemic insect is also deterministic, as is complete removal of a patch of forest that is the only last refuge for an endemic species.

Stochastic extinctions are those that result from chance effects of changes in environmental impact. Such perturbations may thin or partially fragment a population but not eliminate it, leaving it vulnerable to further random impacts. The smaller a population, the greater its vulnerability to further perturbations, and the shorter the interval between such events the more likely a population will be pushed into extinction.

The disturbed landscape, if not too severe to cause deterministic extinction, will set in motion a chain of events that may lead to at least increased extinction risks. Gilpin and Soulé (1986) identify four of these risk areas, as 'extinction vortices': F (inbreeding depression), A (genetic drift), D (fragmentation of the population) and R (lowering of the population makes it vulnerable to further disturbances and further decreases). For insects, with their small size and generally high susceptibility to adverse environmental influences, it is the D and R vortices that are particularly significant, especially in comparison with vertebrates.

In the context of conservation biology, which is a crisis science dealing with vast numbers of species, PVA is a valuable conceptual background, but because of its time and scientific resource demands can only be carried out on a few species. More urgent regarding the biodiversity crisis is the rapid identification of insect-rich areas and those with taxonomically unique faunas or those that generate species. PVA is a complementary topic that should continue to be carried out on selected exemplary species from different ecosystems.

6.2.2 Genetic aspects

Ecological genetics (Ford, 1975) is one of the foundations for insect conservation biology along with habitat and landscape conservation. In particular, Ford's (1975) studies on butterflies not only illustrated the heuristic value of insects for field genetical studies, but also, conversely, led to a recognition of the vital role that genetics can play in conservation biology (Berry, 1971).

High levels of genetic variation are present in most natural populations, although the significance of much of this variability for adaptive processes is unclear. Continuous variation arising from variation among many genes is central to smooth adaptive responses to environmental change. Expression of this variation may be seen in morphological, physiological, behavioural or life-history traits (Brakefield, 1991). Fragmentation and disturbance makes the often highly fluctuating insect populations vulnerable to chance adverse effects of the environment. From a genetic viewpoint, fragmentation also subjects the small populations to the random sampling process of genetic drift. Inbreeding depression (where the small population loses heterozygosity and becomes less adaptive) is probably of greater significance for small, wild populations of vertebrates and for captive populations of insects than for wild populations of insects where environmental impacts are of greater significance.

Polymorphism, whether involving pleiotropy or not, is a fairly common phenomenon in insects as a whole. Cryptic polymorphisms detected by such techniques as gel electrophoresis (for detection of enzyme differences) and DNA technology (for nuclear and mitochondrial DNA identification) for coding and non-coding regions of the genome, are being intensively investigated. Such techniques can indicate how much gene flow there is between populations, and whether the populations are being increasingly fragmented through disturbance to the biotopes and landscapes. McKechnie, Ehrlich and White (1975) showed that, for the rare checkerspot butterfly (*Euphydryas editha*) in California, seven of the eight enzyme genetic loci show much more similarity across populations than can be attributed to present-day patterns of gene flow. Either selection is influencing the seven polymorphisms in similar ways in all the populations, or the patterns are relics of populations that were more continuous in the past.

Species that occur at the edges of their ranges, where physical and biotic conditions are marginal, are particularly susceptible to the genetic effects of small and fluctuating populations as well as fragmentation and extinction of their populations from environmental events. Such species are also subject to local selection pressures.

If only a small number of individuals of these populations are contributing to the next generation, there will be an increasing loss of genetic variation the smaller the population size. Brakefield (1991) has pointed out that a population of 10 males and 1000 females will lose genetic variation at the same rate as one of 20 males and 20 females. The small number of males will act as a 'bottleneck' because half of the genetic information inherited by the next generation comes from each sex (Figure 6.2). With insects, estimates of population sizes are difficult to obtain, except after intensive autecological studies of the type summarized by Dempster (1975). But some studies have been done both in the laboratory and in the field (Harrison, Murphy and Ehrlich, 1988; Bryant, Meffert and McCommas, 1990). These have been

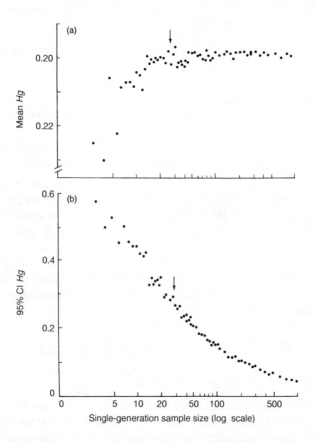

Figure 6.2 Results from simulations involving applying single-generation sampling effects (bottlenecks) to a hypothetical 'total' Isles of Scilly population of the colour-polymorphic spittlebug *Philaenus spumarius*. (a) A mean index of genic diversity (heterozygosity, *Hg*) and (b) 95% confidence interval of *Hg*. Arrows indicate the magnitude of sampling effect yielding a rate of loss of the rare melanic alleles similar to that observed among the smallest islands. (From Brakefield, 1989, 1991.)

reviewed by Brakefield (1991) from an insect conservation biology viewpoint. He has suggested that if a population does not first become extinct for purely ecological reasons (i.e. a stochastic extinction), novel patterns of genetic variation associated with a short bottleneck may provide potential for successful re-establishment and for adaptive responses to environmental change or disturbance. But bottlenecks extending over many generations may, in contrast, be detrimental. With loss of genes, the population has less possibility of adapting to future environmental perturbations.

Bearing in mind the points above, Brakefield (1991) suggests the following components be incorporated into management of insect populations or communities to minimize loss of genetic diversity:

1. Maintain a large population size, while maximizing the proportion of adults contributing to reproduction.
2. Minimize the incidence of bottlenecks in population size.
3. Minimize the duration of bottlenecks.
4. Maintain the movement of individuals and hence gene flow between local populations.
5. Maintain environmental heterogeneity (but not through severe fragmentation) both within and between biotopes across the landscape.

In general, the immediate concerns over loss of habitat automatically compensate for concern for genetic problems. However, in the longer term, there must also be emphasis on genetic variances and on the evolutionary potential. This is still theoretical, and much further research and genetic monitoring is required. From both genetic and the patch dynamic viewpoints, it is most important to maintain as large a population over as wide a geographical range as possible. Such geographical breadth will cover a variety of slightly different biotopes and landscapes, and will also maximize resistance to the impact of global warming.

6.3 NATURAL DISTURBANCES AND PATCH DYNAMICS

6.3.1 Significance of disturbance for insects

Successful selection of suitable hosts is essential for population establishment at a site. These behavioural attributes represent evolutionary adaptations that enhance species persistence within a constantly changing environment and are, therefore, the primary determinants of species responses to disturbance (Schowalter, 1985).

Interaction occurs between adaptive responses to changing environmental conditions with consequences upon the community and ecosystem. Insects vary enormously in their response to disturbance, some insects locating and benefiting from the disturbed patch, while others are destroyed by the disturbance. There will also be variable responses depending on the type of disturbance, its intensity and extent. This in turn will, for example, have an impact upon predators, parasitoids, plant food sources and shelter.

In the context of insect conservation biology, species respond to disturbance in various ways. Dispersal powers towards or away from the disturbance vary, as do survival abilities. In a 'pristine', albeit always dynamic, environment, the various species have already been naturally selected to survive the prevailing conditions. Depending on the severity and extent of the disturbance, and the time-scale involved, species will come and go locally, fragment into metapopulations or even become extinct in that geographical area. For the sake of those species only adapted to narrow tolerances, maintenance of large, pristine biotopes, the larger the better, is essential. Hence the particular concern for preservation of large areas of high endemism in the tropics, where the physiological tolerances of insects to change are generally low.

Although the geographical range of a species may be wide, it may not be continuously distributed across that range. The range, in effect, is a patchwork of populations. As the years pass, each of these populations will be hit by an adverse impact, whether from extremes of weather or from a biotic impact such as a disease epidemic. Very small populations will also suffer from genetic restrictions associated with a very low number of interbreeding individuals. This reiterates some of the conclusions of the

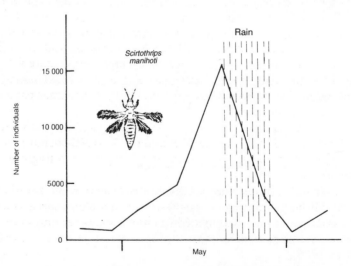

Figure 6.3 A drop in population level of thrips (*Scirtothrips manihoti*) on cassava in central Brazil during several days of tropical downpour. Dead individuals were readily encountered adhering to the water droplets. Figures refer to the numbers of individuals per 20 plants. (From Samways,1989c.)

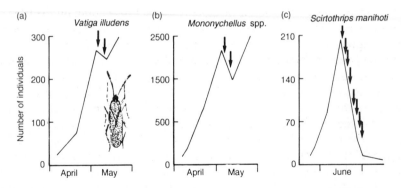

Figure 6.4 The reduction in population levels of a lace bug (a), a spider mites (b) and thrips (c) on cassava in central Brazil at the onset of heavy morning dews (arrowed). Figures refer to numbers of individuals per 20 plants. (From Samways, 1989c.)

SLOPP debate mentioned in the last chapter: many small subpopulations across a wide range, with intrinsically variable levels of gene flow, is perhaps the best species survival pattern if a large continuous matrix is not possible.

6.3.2 Inclement weather

Severe weather conditions, from cold snaps in the temperate regions, to rain storms and cyclones in the tropics, sometimes have visibly dramatic effects upon patches of vegetation. Similarly, heavy tropical rain (Figure 6.3) and dew (Figure 6.4) frequently cause a surface tension trap for small insects, and prolonged cool and damp weather can inhibit flight or encourage pathogens that reduce insect populations.

Inclement conditions that are exceptional in being sudden and severe or prolonged, such as long cool periods or extremely dry ones, can have a major impact upon insect populations, sometimes causing them to fragment into metapopulations.

Some insect populations recover remarkably quickly from disturbances that subside fairly quickly. Figure 6.5 shows how an odonatan assemblage recovered within a year in response to flood levels that occur only every 100 years or so.

6.3.3 Burning

Natural burns on the landscape from lightning strikes are a widespread phenomenon, particularly in savanna areas such as parts of Australia (Gill,

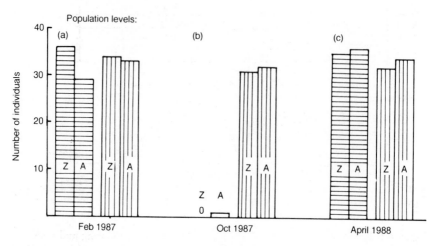

Figure 6.5 Changes in population levels of damselfly (Z = Zygoptera) (*Pseudagrion kersteni, P. salisburyense* and *Platycypha caligata*) and dragonfly (A = Anisoptera) (*Trithemis arteriosa, T. dorsalis, Notogomphus praetorius* and *Paragomphus cognatus*) larvae in a stream (horizontal bars) and a pond (vertical bars) before (a), within 2 days (b) and several months after (c) the torrential downpour and devastating floods in Natal during late September 1987. (From Samways, 1989c.)

Groves and Noble, 1981) and Africa. Burning maintains the savanna as grass-land, preventing succession to natural scrub (Tainton and Mentis, 1984).

Campbell and Tainton (1981) found that spatial and temporal variability within Australian eucalypt forests mask and make it impossible to fully assess the impact of fire on the soil and litter invertebrates. They noted that recovery of the community was generally quite rapid, and that no invertebrate order that they investigated appeared to have an adaptive response solely to survive fires. The invertebrate orders least affected by fire showed high mobility or adaptations to conserve water and resist high ambient temperatures in seasonally dry biotopes. These adaptive responses shown by this Australian invertebrate fauna were already inherent in the community before the fire. Recolonization after the fire was also associated with the dispersal ability of the insects.

Nevertheless, fire does have transient influences on insects. Fire initially deprives grasshoppers of food (Gandar, 1982). Elevated post-burn soil temperatures, through increased insolation, allow some grasshopper nymphs to emerge earlier than usual, which, coupled with more nutritious vegetative regrowth, progressively increases grasshopper populations (Warren, Scifres and Teel, 1987). Additionally, profuse vegetation regrowth can also cause enhanced maturation of females, leading to an increase in the numbers of eggs laid. The increased post-burn insolation of the soil favours simultaneous

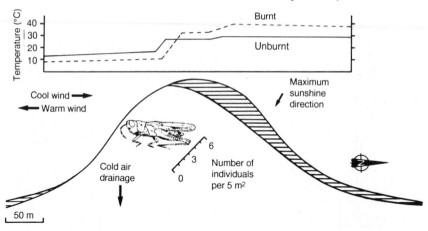

Figure 6.6 Temperature variation of the soil surface at midday during mid-winter on the burnt and unburnt hillocks in the Natal Drakensberg, South Africa (2000 m a.s.l.). Differences were most extreme on the burnt terrain. Maximum insolation was from the north, which was also subject to warm daytime winds. Cool night-time winds from the south and cold air drainage were also significant. These mesoclimatic influences affected grasshopper abundance: *Dnopherula* sp. nr *cruciata, Machaeridia conspersa, Faureia rosea, Anthermus granulosus, Acorypha ferrifer* and *Acanthacris ruficornis*. (From Samways, 1989c.)

hatching of eggs, leading to relatively stable year-to-year population levels.

When burn is interrelated with topography it can reduce grasshopper population levels by half. Although many species are adapted to burn conditions, the reduced grass cover accentuates cold-air drainage at night causing increased mortality (Samways, 1989c, 1990b) (Figure 6.6.).

In contrast, some beetles even depend on fire and have a highly developed tactic response to smoke to locate burnt trees in which to lay their eggs (Mitchell and Martin, 1980).

From an insect conservation biology viewpoint, fire, although influential on many insect species, is not generally harmful. On the contrary, some ecosystems, with their inherent insect fauna, depend on fire for their maintenance. Localized non-intensive fires in high fire-risk areas appear neutral or beneficial overall for insect communities. In contrast, runaway fires and those that cause complete destruction of indigenous forest in areas not naturally subject to such major impacts are detrimental to insect conservation. Fire management to simulate natural fires where the landscape has been fragmented by roads is an important management practice for maintaining certain wilderness and savanna areas.

6.3.4 Compound and cyclical effects

Disturbance to the landscape can often have a compound effect on the biota. Insect-induced vegetation mortality can open the forest canopy and create fuel accumulations that predispose an area of forest to disturbance (Schowalter, 1985). The nature of the ensuing disturbance determines the future growth and migration of insect populations and the future rate and duration of localized community development. This means that the pattern of insect population interactions with the disturbance is a major determinant of patch dynamics.

Schowalter (1986) and Schowalter, Coulson and Crossley (1981) use as an example the North American pine forests. Fires during summer droughts in the fire-tolerant pine forests of the south-eastern USA cause the forest to keep its characteristic composition. But the southern pine beetle (*Dendroctonus frontalis*) is an occasional herbivore of pine trees over 30 years old which are stressed or damaged and infected with 'blue-staining fungi'. Normally, the beetles cannot overcome the oleoresin defences of the younger pines. Beetle populations build up around these damaged trees, causing a wave of pine mortality. Forest regeneration then occurs. In the absence of subsequent disturbance, beetle-induced mortality of the dominating pines accelerates the transition to fire-intolerant hardwood forest, with the youngest hardwoods in the immediate wake of the mortality wave and progressively older hardwoods extending away from the wave. However, these forests are subject to frequent disturbance, with severe ice and wind storms damaging exposed vegetation behind the wave and along its edge, increasing vulnerability to further damage. Then during log degradation, there is a strong chance that lightning will eventually ignite a standing pine stand. Fire, fuelled by abundant woody litter, probably ends the advance of the insect-induced mortality wave but provides new beetle refuges in fire-scarred pines. Fire also destroys the hardwoods, encouraging pine forest again, until after 30 years the insect-induced mortality begins again (Figure 6.7).

Studies of interactions such as this are extremely valuable for insect conservation biologists, emphasizing the time and space dynamics of communities, and how the interrelatedness of the plants, insects and abiotic factors must all be taken into consideration.

Keeping interactions intact, even simply the plant–insect association, does not alone guarantee survival. J.A. Thomas (1989) cites the case of the county of Suffolk in England, which has lost 42% of its butterfly species this century, yet none of their food plants have disappeared. The large blue butterfly (*Maculinea arion*) until 1979, when it became extinct in Britain, occurred in several sites in the south-western part of the country in mutualistic association with the ant *Myrmica sabuleti*. Many of the sites were lost to agriculture,

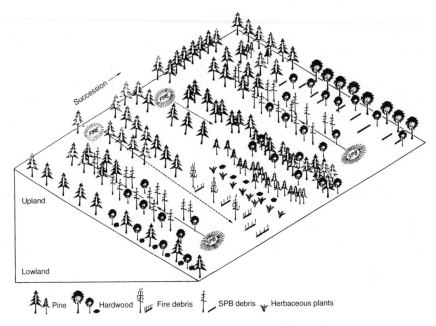

Figure 6.7 Representation of the south-eastern coniferous forest (upland and lowland) illustrating interactions between *Dendroctonus frontalis* (southern pine beetle, SPB) and fire. Successional transformation resulting from fire and *D. frontalis* extend from left to right: dotted arrows indicate the direction of movement. Fire, a regular feature of the generally dry, well-drained uplands, invades generally moist, poorly drained lowlands, where drought or *D. frontalis* creates favourable conditions. *Dendroctonus frontalis* depends upon fire for regeneration of pine stands. The hardwood climax forest (far right lowland) results from freedom from fire and can be reduced by fire. (From Schowalter, Coulson and Crossley, 1981.)

urbanization or quarrying, or were undergrazed by wild and domestic herbivores, causing a large reduction in the ant, which in turn caused the demise of the butterfly (J.A. Thomas, 1980).

6.3.5 Mutualistic interaction resilience

The continuance of the biotope, and the resilience of mutualistic relationships, vary considerably; some of the most fragile are of conservation concern. Contrast the *Maculinea–Myrmica* mutualism with a considerably more robust interaction. The African pugnacious ant, *Anoplolepis custodiens*, is mutualistic with the citrus mealybug (*Planococcus citri*). The fungus *Cladosporium oxysporum*, along with other natural enemies, can reduce populations of this widespread, polyphagous and abundant mealybug. *A. custodiens* attends and protects *P. citri,* but cannot totally stop the impact of a *C. oxysporum* epizootic reducing the mealybug population. The ant appears to groom the mealybug free of potentially lethal fungal spores. The ant continues to protect a few selected groups of mealybug, and in the meantime hedges its bets by attending the cotton aphid (*Aphis gossypii*). The aphid population also soon succumbs to the disease, and the ant population falls as a result of reduced food resources. However, the point is that the interaction is maintained between the ant and the mealybug at a low level, with each gradually building up again, keeping *C. oxysporum* at bay, until the next cycle (Samways, 1983a) (Figure 6.8).

The butterfly–ant and the mealybug or aphid–ant mutualism varied in their robustness in several respects. Both the large blue butterfly and its ant mutualist were highly susceptible to the disturbance, while the mealybug and aphid, although severely reduced in numbers, were resilient enough eventu-

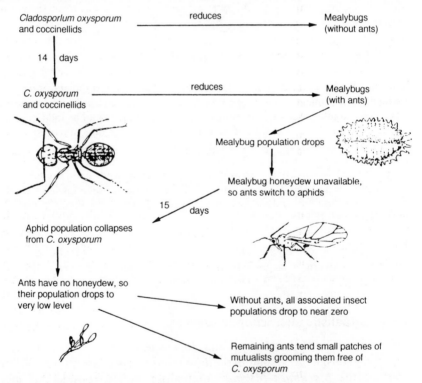

Figure 6.8 A remarkably resilient mutualistic relationship between the ant *Anoplolepis custodiens* and its mutualists *Planococcus citri* and *Aphis gossypii* in the face of fungal epizootics by *Cladosporium oxysporum* on guava trees. (From Samways, 1990c.)

ally to overcome the disturbance. Further, the large blue butterfly had a restricted distributional range in Britain, which was finally whittled down to nothing. It is probably significant that these British populations were also on the margin of the butterfly's overall range, where conditions were possibly suboptimal for one link in the mutualistic chain. In short, rigid or non-resilient interactions are the first to succumb to disturbances if confined to a patch, whether it be a reserve or not, that is susceptible to some other adverse disturbance of major importance, extinction is the likely outcome.

6.3.6 Implications from models

Landscape ecology is a young science, and modelling of landscapes has a long way to go before it can be prescribed for insect conservation biology. The disturbance landscape models of Turner *et al.* (1989) are the basis of experimentation. They found that susceptible biotopes that occupy less than 50% of the landscape are sensitive to disturbance frequency but show little response to changes in disturbance intensity. Susceptible biotopes that occupy more than 60% of the landscape are sensitive to disturbance intensity and less sensitive to disturbance frequency. These dominant biotopes are also easily fragmented by disturbances of moderate intensity and low frequency. This means that patches of rare communities are more susceptible to frequent disturbances than to intense disturbances. A locally intense disturbance may eliminate a part of the biotope or patch but have little effect on the persistence of that biotope in the landscape. In contrast, a large number of low-intensity disturbances over a large region could reduce or eliminate that type of biotope.

Biotopes that are common may be easily fragmented by disturbances of only low to moderate intensity. Intermediate levels of disturbance intensity and frequency create, according to Turner *et al.*'s (1989) models, greater patchiness in landscapes that were dominated by disturbance-prone biotopes. For example, large tracts of forest may be fragmented by disturbances of relatively low frequency. Structural changes associated with this fragmentation, such as the increased number of edges, have important implications for the susceptibility to other disturbances and for the distribution and abundance of plant species, and their associated insect species. These effects are likely also to be suppressed by the feedback interactions between the insect community and the plants.

Turner *et al.* (1989) imply that, if the biotope type is rare, management should focus on the frequency of disturbance. Disturbances with low frequencies will have little impact, even at high disturbance intensities, because there is insufficient landscape connectivity. In short, new disturbances will tend to be contained by the landscape structure, but if disturbances are frequent landscape structure will change. On the other hand, if the biotope type is common, management must consider both frequency and intensity. As biotope and landscape quality is the main dictate for insect preservation, these models are well worth testing on real insect communities. Such empirical findings would be of general value, although their local value will be gradually overridden by the increasing intensity of global climatic changes.

Theoretical ecological studies have indicated in general terms that simple ecosystems, with fewer energy transfer interactions, are more robust than more complex systems. Framing this in conservation terms means that more complex systems need more protection from disturbance than the simpler ones (but c.f. caves, p.93). Complex systems contain proportionately more rare species, more genetic uniqueness and more rare species interactions than simpler systems. Impacts on complex systems are also more likely to affect a variety of insect species that are rare because of highly restricted distributions and/or very specific behavioural or ecological requirements.

6.4 FROM ADVERSITY AGRICULTURE TO AGROECOLOGY

6.4.1 Soil loss and fossil fuels

Any form of nature conservation cannot deny human needs for food. A starting premise is that natural vegetation has and will continue to give way to agricultural development. The urgent emphasis today is for managed development, rather than destructive unplanned expansion. There cannot be total disregard for the natural cornucopia of living material or disregard for the soil fabric. Land is finite. In fact it is less than that, with South Africa alone losing 300 million tonnes of topsoil annually. This brings to mind the aphorism that there is only about 30 cm of soil between life and death of the planet. Worldwide this precious layer is being lost. Even the USA is losing topsoil 18 times faster than it is being replaced (Pimentel *et al.*, 1989). We are living off our capital.

The fossil energy being put into technological agriculture has increased enormously. An estimated 17% of the annual fossil energy consumption of the USA is used to supply the country with its food and fibre needs. About 1100 litres of oil equivalents is required to produce a hectare of a crop such as maize. The energy input for nitrogen fertilizer alone is now greater than the total energy inputs for growing maize in 1945 – about a 20-fold increase in the amount of nitrogen fertilizer versus a threefold increase in maize yield (Pimentel *et al.*, 1989).

These impacts have a major detrimental effect on soil invertebrates, which are now recognized as being important influences on process-level dynamics in agroecosystems (Crossley, Coleman and Hendrix, 1989). Most of the 200 000 species of plant and animal species that occur in the USA are involved in some way in agricultural production by performing many essential functions including degrading wastes, recycling nutrients, protecting crops and livestock from pest attack (particularly natural enemy activity), pollinating crops, conserving soil and water resources, and preserving genetic material for crop and livestock breeding.

In the USA, despite all pest controls, which includes use of about 318 000 tonnes of pesticide annually, about 37% of crops ($50 billion worth) is lost to pests (Pimentel, 1986). Although this heavy pesticide usage has reduced damage caused by some pests, there has been no overall reduction in crop losses from pests. In fact, crop losses caused by insects have increased nearly twofold (from 7% to about 13% of crop yields) from 1945 to the present. This has occurred despite a more than 10-fold increase in insecticide use during the same period (Pimentel *et al.*, 1989).

There will inevitably be a move towards less reliance on agrochemicals, with greater dependence on green products (Lisansky, Robinson and Coombs, 1991). Although the yields of organically grown crops are generally less per land area than those dependent on high agrochemical input, the gross economic margins, at least in the UK are higher. Thus green growing will tend to go hand in hand with greater preservation and promotion of insect diversity.

6.4.2 Landscape degradation

Landscape alteration and the impact on insect populations are recognized much more today than in the past. This refers not only to adverse impacts on rare species, but also to explosions in insect populations to pest proportions. This in turn may lead to unnecessary and inappropriate control measures. The South African Karoo caterpillar (*Loxostege frustalis*) sporadically reaches severely high proportions in the arid and semiarid Karoo region (Annecke and Moran, 1977). Over a period of 10 years, 12 parasitoids were imported and mass-reared in millions, but they never became established or had any impact upon the pest. A change in research direction illustrated that the real reason for the species reaching high, destructive levels was that the landscape had been altered to increased scrub-bush density through overgrazing by domestic animals. This favoured aspects of the insect's biology at the expense of the natural grasses and herbs. Appropriate pasture management now aims at reducing the status of the pest, but circumstantial evidence indicates that damage was done to other rare and localized insect species during the initial process of landscape degradation.

6.4.3 Tenets of sustainable agriculture

Agriculture, as well as urbanization, has increased in intensity, increasingly 'dominated' natural processes and increased in expansiveness. The fragmentation has increased in scale, in keeping with the convenience of larger scale mechanized tillage and harvesting. There is now a trend for this landscape monotonizing to be slowing down in some areas. In Britain, government economic incentives have encouraged a switch from excessively intense agricultural production to preservation of parts of the farmed landscape.

Not all agricultural societies have employed highly mechanized methods, and some have even resisted high-energy input agriculture. The Amish community of the USA has rotated crops for pest control alongside maintenance of landscape diversity for over two centuries (Stinner, Paoletti and Stinner, 1989). Even some modern methods based on sound agroecological premises have proved to be beneficial in many respects. *Shizen* farming of rice in Japan employs differences in management practice compared with conventional methods. Planting techniques, plant density, irrigation practices, fertility sources and quantities, insect disease and weed management tactics are modified. Rice yields from *Shizen* and traditional farming are similar, yet pest attack on plants early in growth is reduced (Andow and Hidaka, 1989).

Sustainable agriculture encompasses three tenets, outlined by the *World Conservation Strategy* (IUCN, 1980), which also applies to natural ecosystems:

1. Maintenance of essential ecological processes and life-support systems;
2. Preservation of genetic diversity;
3. Sustainable utilization of species and ecosystems.

In the update to this publication, *Caring for the Earth* (IUCN, 1991), sustainability is emphasized even more strongly, with the directive to build a sustainable society. Such a society must continue forever, without depletion of the world's resources. This is based on nine principles:

1. Respect and care for the community of life.
2. Improve the quality of human life.
3. Conserve the earth's vitality and diversity.
4. Minimize the depletion of non-renewable resources.
5. Keep within the earth's carrying capacity.
6. Change personal attitudes and practices.
7. Enable communities to care for their own environments.
8. Provide a national framework for integrating development and conservation.
9. Create a global alliance.

The action plan arising from these principles is to develop new strategies for sustainable living.

6.4.4 Insect conservation biology and sustainable agriculture

Insect conservation biology contributes to the *Caring for the Earth* action plan in two ways. Firstly, it scientifically underpins the preservation of as many insects as possible through landscape conservation. Secondly, it encourages the use of natural predation, parasitism and insect pathogens by appropriate landscape and crop management. Encouraging indigenous natural enemies to perform in the sustainable context poses little or no threat to the conservation of the insect fauna. This contrasts with some of the risks associated with the introduction of exotic biocontrol agents, discussed further in Chapter 8.

Agroecology classifies and studies agricultural systems from an ecological and socioeconomic perspective (Altieri, 1989; House and Brust, 1989). It is a scientific approach used to study, diagnose and propose alternative low management of agroecosystems. Solving the sustainability problem of agriculture is the primary aim of agroecology (Carroll, Vandermeer and Rossett, 1990).

Insect conservation features strongly in this approach in all areas of the ecosystem: soil maintenance, pollination and maximization of natural enemies. Techniques involve planting of hedges in temperate (Nazzi, Paoletti and Lorenzoni, 1989) and tropical lands (Samways, 1992b), which then act as reservoirs for natural enemies. Fields of cassava (*Manihot esculenta*) in Brazil, when extensive, are at risk from severe pests such as the cassava hornworm (*Erinnyis ello*), the lace bug (*Vatiga illudens)* and the cassava mite (*Mononychellus* sp.). In small fields surrounded by hedges and natural vegetation, these arthropods are rarely pests. There is high predation and parasitism from natural enemies, which rapidly move into the crop, causing extremely high mortality among their prey (Samways, 1979). Interestingly, this interaction was not influenced by the level of cyanogenic glycosides in the cassava cultivars, indicating that the landscape pattern was a more important determinant of level of herbivore suppression than aspects of plant chemical composition.

6.4.5 Conservation headlands

Use of conservation headlands (borders of unsprayed crops around the edge of a crop field) (Sotherton, Boatman and Rands, 1989), encouragement of mixed flowering systems as nectar and refuge areas for natural enemies, reduced herbicide and fertilizer usage, as well as mixed cropping, are also conservation procedures. These methods are anthropocentric and are designed to achieve sustainable agricultural output for man's benefit, although conservation headlands can preserve fauna without necessarily any agricultural gain.

These headlands are 6-m-wide edges around cereal fields that receive reduced and selective pesticide inputs (Dover, 1991). The system is under continuous development in that different methods of managing this 6-m headland are being tried. It is not simply a case of leaving this strip free of treatment, because factors such as weed invasion and control must be built into the management formula. Although there is a 5–10% crop yield loss when employing conservation headlands, they provide extra host plant and nectar resources which benefit many edge species.

Dover (1991) has demonstrated that the total populations of various edge species of butterflies in the Hesperiidae, Lycaenidae, Nymphalidae, Pieridae and Satyridae are overall about twice as abundant in conservation headlands than in fully sprayed headlands. These conservation headlands increase the population levels of many organisms besides butterflies. But for the butterflies at least, none are to date truly threatened species. The aim of this conservation measure is to promote an increase in wild biomass rather than directly to save species from extinction. This is an important perspective, and conservation headlands may, in areas of high endemism in the tropics, and on islands, as well as in the northern and southern hemisphere, have an application in directly preventing species extinction.

6.4.6 Conservation corridors and patches

With the emergence of the concept of sustainability coupled with landscape ecology, attempts are now being made to analyse the spatial aspects of plant patterning associated with agroecosystems for energy conservative methods of pest control. Uncultivated or sustainably managed networks of corridors can have beneficial functions (Forman and Baudry, 1984), including better control of insect pests in comparison with bare ground separating plots (Stinner, Rabb and Bradley, 1974; Rabb, Stinner and van den Bosch, 1976). But there is considerable experimental work to be done to elucidate the nuances of sustainable methodology. An experimental study by Kemp and Barrett (1989) showed that soybean fields separated by grassy corridors contained fewer Mexican bean beetles (*Epilachna varivestis*) than plots separated by uncultivated corridors. The fungal pathogen *Nomuraea rileyi* infected higher proportions of the pest moth caterpillar green cloverworm (*Plathypena scabra*) in plots divided by grassy corridors. Other types of corridor, including uncultivated ones, were investigated, and uncultivated areas did not 'funnel' predators into the crop as hypothesized. Nevertheless, overall, Kemp and Barrett (1989) suggest that management of uncultivated corridors within the landscape is an important means of regulating insect pests.

Predator conservation islands can be extremely important in ecological and economic terms. One of the reasons that the coccinellid biocontrol agent

Chilocorus nigritus is so successful is that it roosts in clumps of the giant bamboo, *Dendrocalamus giganteus*, during winter and moves into citrus orchards to control red scale (*Aonidiella aurantii*) during spring and summer (Samways 1984a; Hattingh and Samways, 1991). Further, this shuttling behaviour is particularly important, as red scale is highly resistant to organophosphate insecticides, making a biocontrol safety net against scale outbreaks especially valuable.

Galecka (1991) describes the interesting case of Polish pino-quercetum and birch-aspen woodland patches being winter refugia for as many as 24 species of coccinellid, which disperse into adjacent crop fields in the summer to suppress aphid populations.

Conservation patches, whether remnant or introduced, would be expected to lose certain species over time frames of many years, especially the smaller the patch. Turin and den Boer (1988) found that in smaller patches of woodland carabid beetles with low powers of dispersal become rarer and increasingly threatened in comparison with more widespread, mobile species. Mader (1984) has proposed that forest patches need to be between about 2 and 5 ha to maintain the sylvan carabid assemblage intact. For wandering spiders, the area is likely to be nearer 10 ha.

6.4.7 Mixed-crop patches

Besides using agricultural conservation corridors, there is also the potential for modifying the within-crop structure to improve natural enemy activity. Traditional small patches of highly mixed crops surrounded by native tropical vegetation are generally pest-free and rich in native insect diversity (Figure 6.9).

The term 'pest' is an economic concept used to describe species that cause crop losses and thus monetary losses. Mixed-crop patches can only suffer limited damage, as it is unlikely that any particular pest would seriously affect more than one type of crop. There is a dynamic equilibrium between the potential pest (the host) and its natural enemies. It cannot be said that there is a 'balance of nature'. Such a concept does not exist demographically, and certainly not economically. Hosts and their natural enemies are in constant spatial and temporal flux. When the potential pest demographically escapes the natural enemy suppression, then it may inflict economic damage and therefore be an actual pest.

In localized, low fossil-energy input agricultural patches, insect herbivores (as well as mites) are continually suppressed in two main ways. First, the individual crop plant is locally rare, i.e. it is outnumbered by many other plant species, both naturally occurring and planted. This so-called low apparency means that the pest must first find its crop host. Secondly, there is ample opportunity for the variety of natural enemies to utilize the variable

Figure 6.9 A small patch of mixed crops surrounded by natural vegetation in Indonesia. The crops were relatively pest-free and native insect species richness was high.

plant architecture for optimal foraging. Natural enemies also have to contend with low apparency of their hosts (the potential pest), but their searching abilities, sharpened by natural selection for interspecific competitive advantage, appear to overcome the low apparency. Further, the age structure of the host is subject to differential attack by different natural enemies, which together bring the pest below the economic threshold (Samways, 1988a).

6.4.8 Crop polycultures for insect predator conservation

Within the context of sustainable agriculture, there is still a long way to go before even beginning to understand the complexities of interactions in polycultures. As Andow (1991a) puts it, the number of potential ecological interactions between plants, arthropod herbivores and arthropod natural enemies, and the possible evolutionary responses of each population to any of the others, creates a veritable Gordian knot of complexity. Andow (1991a) calculates that a relatively simple ecosystem of two plant species, six herbivore species, and six natural enemy species has 91 potential two-way and 364 potential three-way ecological interactions, and at least an equal number of possible evolutionary responses. These interactions make generalizations on methodology difficult and unrealistic.

In a survey of various types of monocultures and polycultures, Andow (1991b) found that marginal benefits accrue from arthropod pest control in polycultures when those polycultures have lower pest populations than monocultures. But no benefit occurs when polycultures have similar or higher pest populations. There appears to be an interaction not just between herbivores and their natural enemies and plant architecture, but also between the plants themselves, limiting the ability of the crop to compensate for pest injury or crop tolerance. This further emphasizes that pest management practices must be individually tailored to the particular crop production and conservation goals. Such goals need to be well-determined beforehand. There is no conflict between natural control of pests by carefully modified landscapes and between maintenance of biodiversity. The move away from highly extensive monocultures and widespread reliance on high-yielding crops with high fossil fuel requirements and the tendency for more integrated landscape management is acceptable to pest managers and conservationists alike. But much research is still required at various spatial and temporal scales on various crops in various regions to ascertain the optimal tactics.

6.5 BENEFICIAL ASPECTS OF THE AGRICULTURAL LANDSCAPE ON INSECT CONSERVATION

6.5.1 Woodlands and forests

Primary and old-growth woodlands or forests, whether in Europe (Rackham, 1980), North America (Harris, 1984), Australia (New, 1984) or the tropics (Jacobs, 1988), are unquestionably major climax growth areas whose fragmentation into patches is to be avoided as it is damaging to interior species. For Panamanian forests the greatest changes occur between 2.5 and 15 m into the forest, and as far as the vegetation is concerned the exterior conditions prevail 15–25 m into the forest (Williams-Linera, 1990). In view of the environmental sensitivity of tropical insect species to environmental stress (Parsons, 1989), the interior species presumably would need at least 25 m of edge to benefit from the normal range of conditions to which they are adapted. This excludes the possibility that the edge itself remains intact and does not move interiorly with soil run-off, typhoon damage, etc.

With probably at least 95% of the British ancient wildwood gone, and 93% of the Madagascan and 95% of the coastal Brazilian forest strongly disturbed or removed, only remnants remain some areas. To save species and maximize natural predation, agriculture in forests must be minimal in extent. Localized clearances, which benefit both edge and interior species, can then be maintained sustainably. Alternatively, there must be rotational, sustainable utilization to maintain the interior environment for forest and woodland-interior insect species.

6.5.2 Hedges

Hedges (Pollard, Hooper and Moore, 1974) are a characteristic disturbance feature of the north-temperate landscape. Ancient hedges are rich in fauna and flora, including insects (Lewis, 1969a), as further indicated by being a

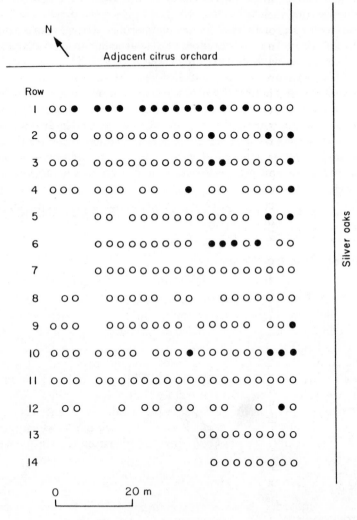

Figure 6.10 An orchard of young citrus trees illustrating the accumulation of adult citrus psylla *(Trioza erytreae)*. The psylla population builds up in trees nearest to an adjacent old citrus orchard and a row of silver oaks (*Grevillea robusta*) which act as a windbreak. ● = trees with > 20 adult *T.erytreae*, ○ = trees with < 20 adult *T. erytreae*. (From Samways and Manicom, 1983.)

source of natural enemies in temperate (DeBach, 1974) and tropical lands (Samways, 1979). Their size, direction and composition have an important influence on insect movements (Lewis, 1969b). For some insects there is a definite pattern of movement at the interface between fields and their borders. In the case of the citrus psylla *(Trioza erytreae)*, there is a highly significant logarithmic increase in immigration rate of adults and in the number of new shoots supporting their eggs after the start of the growing season. There is increasing clumping of adults with time, following a power curve, so that the final dispersal pattern is of adults accumulating in the tree rows adjacent to the borders (Samways and Manicom, 1983) (Figure 6.10).

The indications are that hedges are highly variable in their nature and significance for, and impact upon, insect populations. Hedgerows influence not only the aerial fauna but also soil fauna, often with enhanced predation of herbivores and detritivores (Nazzi, Paoletti and Lorenzoni, 1989). Even in Britain, with its high research input, the insects that inhabit or are associated with hedgerows have not been fully listed (Jones, 1991). However, important groups include Lepidoptera, Hymenoptera, Psyllidae, Miridae, Pentatomidae, Chrysomelidae, Carabidae and Coccinellidae.

The value for insects lies partly in hedgerow's varied structural and plant components, although these vary across Britain and Europe. Jones (1991) identifies four structural components:

1. **Hedge body.** A more or less continuous line of shrubs and bushes, which, when of considerable width, could almost be considered as an 'interior' as well as an edge biotope for many species. In some respects this is equivalent to the mantel, the shrubby border to the forest edge (Forman and Godron, 1986).
2. **Hedgerow trees.** These are virtually all edge as they generally stand exposed on both sides.
3. **Hedgerow verge.** This is the strip of land either side of a hedge relatively undisturbed by agricultural activities, except grazing stock. Here perennial herbs may flourish, and it is equivalent to the saum of the forest edge (Forman and Godron, 1986).
4. **Ditches.** Although they have an adverse effect by draining wetlands, ditches provide temporary or permanent biotope for aquatic and semi-aquatic insects such as Odonata and Heteroptera (Figure 6.11). Many hedgerows do not have this ditch component, and instead border an ancient road, which acts like a ditch in being much lower than the bordering hedgerows.

Also of great importance is the fact that hedgerows provide shelter, food plants (e.g. elm suckers provide food for the white-letter hairstreak butterfly, *Strymonidia w-album*) and nectar sources, particularly the Umbelliferae, which are attractive to the Syrphidae and Cerambycidae for example.

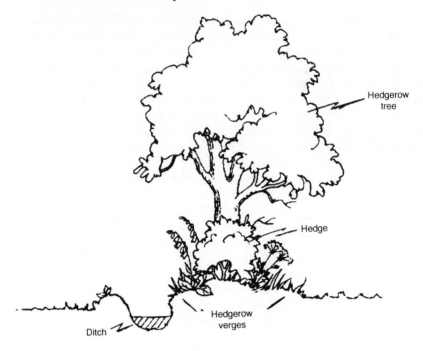

Figure 6.11 Structural components of a common type of English hedgerow. (From Jones, 1991.)

Hedgerows are managed in various ways, but many of the labour-intensive methods have given way to mechanical and chemical methods of cutting and management, which have varying impacts upon the insect fauna (Jones, 1991).

Unquestionably, hedgerows provide a rich and unexplored area for the insect conservation biologist. Particularly fertile areas of research include the influence of hedge intersections and hedge openings on insect community patterns and behaviour, i.e. connectivity and permeability in landscape terminology. Also, an almost unexplored field is the value of hedges in tropical areas, not only for agricultural reasons such as soil conservation and biocontrol agent refugia, but also for conservation of non-commercially important insects, as well as natural predators.

6.5.3 Terracing

Although mountainous areas often provide refuge for biota as non-viable agricultural or urban development areas, some ancient human cultures in Peru, Israel, Afghanistan and Nepal have nevertheless utilized such areas for

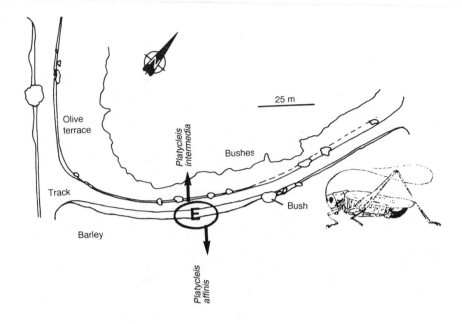

Figure 6.12 The importance of an ecotonal strip corridor of mixed grass, herbs and bushes (E) adjacent to a dirt track at the base of the wall of an olive terrace for two species of bush cricket, *Platycleis affinis* and *P. intermedia*, in southern France. The nymphs matured mixed together throughout this warm, open vegetation, prior to establishing in their adult biotopes on maturing. *P. affinis* moved partially into the barley, while *P. intermedia* moved to the bushes, from where it broadcasts its calling song at night. (From Samways, 1989c.)

protective and military defensive reasons. Terracing is associated with hilly or mountainous terrain, and, by virtue of the walls and banks, often provides refugia, larval food plants, nectar sources and thermal warmth at night that encourage certain insect fauna. Few studies have been conducted, but olive terraces in southern France provide distinct and important biotopes for some highly localized tettigoniids (Samways, 1976, 1977b). The nymphs of two species, *Platycleis affinis* and *P. intermedia*, mature together in the ecotonal vegetation along dirt tracks at or near the terrace bases. On maturing they move to their specific adult biotopes (Figure 6.12). Not only do the bushes, herbs and grasses provide food, but the warm stones at night encourage acoustic and mating behaviour. Interestingly, warm stones also stimulate rock basking and hence foraging behaviour in dragonflies (McGeoch and Samways, 1991).

Even in lowland areas, rice paddies are often terraced to some extent and provide refuge for many aquatic insects. Indeed, the first dragonfly reserve in Japan was converted from rice paddies (Moore, 1987b).

6.5.4 Grazing and mowing

Pastoralism has been an activity of man from the earliest of times. In some areas, such as the central African savanna, grazing both by domestic cattle and by indigenous megaherbivores has continued side by side for centuries. Many grassland insect species have adapted and diversified under the long prehistory and history of grassland savanna of central Africa. The area has been considered a 'species dynamo' (Jago, 1973) for acridids, which show much genetic plasticity and species diversity.

The negative effects of overgrazing by domestic animals are well known, particularly their increase in recent years (Goudie, 1989). However, the converse, particularly in temperate areas, has conservation significance for many invertebrates, especially insects. If grass is left ungrazed or unmowed, food plants can become shaded out. The pearl-bordered fritillary (*Clossiana euphrosyne*) has a strong preference for small violet plants that have sprung up in a warm microclimate (Thomas and Snazell, 1989), yet the small pearl-bordered fritillary (*C. selene*) lays on the same violet plants a year or two later when the plants are larger and surrounded by lusher growth (Fry and Lonsdale, 1991). Indeed, various insect species from various groups, such as the Coleoptera and Heteroptera (Morris, 1971), Lepidoptera (Dempster, 1971; Thomas, 1983a; Thomas *et al.*, 1986), Orthoptera (Bei-Bienko, 1970; Guéguen-Genest and Guéguen, 1987; Marshall and Haes, 1988) and ants (Samways, 1983a), depend on some regular disturbance to maintain their population numbers.

It is the major disturbances such as ploughing, heavy grazing, fertilizing plots, vertebrate grazing, especially by goats, and recreational pressure that contribute to declining population levels, and loss of the local insect populations (Fry and Lonsdale, 1991). Most of the detailed studies listed above were based on considerable research, and indicate that for preservation of the greatest number of species a spatially variable management regimen including rotation mowing and restoration activities such as increasing flowering herbs, is necessary. However, if the aim is to conserve a population of a particular species, then management will depend on specific pointers from autecological studies.

Grass monocultures for improved grazing of cattle in southern Africa impoverishes the orthopteran fauna. Similar practices in Brazil, although also reducing orthopteran diversity, nevertheless cause a hemipteran, the chinch bug (*Blissus leucopterus*), to reach damaging levels. There is unquestionably a need for more research on rehabilitation of grasslands in tropical areas that would benefit insects and the biome in general.

6.5.5 Water courses, ponds and farm dams

Draining of wetlands usually provides fertile agricultural land while leaving conservation remnants of wetland that have to be constantly managed to maintain the water table, as is the case with Woodwalton Fen in England (Duffey, 1971). Where management of this sort does not occur, the remnant usually goes through a dry-land vegetation succession. Drainage channels in these high-lying refugia may be the last area for certain species that originally flourished in the former extensive wetlands. This contrasts with the lower-lying channels, running through agricultural land, which tend to be species depauperate.

Ponds are an important ornamental landscape feature and provide refuge for many insects, not necessarily aquatic, but for moist-land edge species too. Landscaping of the pond is crucial. The tidying up for aesthetics and to improve amenities of Richmond Park, Greater London, has contributed to a 50% decrease in dragonfly species over the past 40 years, a similar pattern to that of Bookham Common, Surrey, UK (Fry and Lonsdale, 1991) (Figure 4.11).

In arid countries, the creation of water bodies has great attraction for many insects. Permanency and management play an important role. For example, artificially created temporary pools, whether for wild or domestic vertebrates, encourage a dragonfly fauna, albeit species poor and of eurytopic species, that otherwise would not be there. However, small farm dams permanently fed by streams, can be extremely rich insect refugia especially where vegetation is left to regenerate naturally (Samways, 1989f). There is, in addition, an interaction with elevation. At all altitudes, most, if not all, dragonfly species occurred at least in dams. Only at the higher elevations (1501–1800 m above sea level) were all the species found in at least natural biotopes. At the middle altitudes (901–1200 m above sea level), 78% were recorded only from dams, with similar but less extreme results at even lower altitudes. In short, high in the mountains where rain catchment was maximal, extra pools of water were of little significance, but lower down, where permanent pools were scarce 'introduced patches', they led to a more intensive distributional pattern of many otherwise quite localized species. However, such dams are not to be confused with large-scale river impoundments, which may have major compound effects upon biota above and below the dam wall for many kilometres.

With such a dramatic loss of wetlands worldwide, particularly low-lying coastal ones, it is essential to focus upon them as much conservation research as possible. One reason is that loss of natural wetlands means impoverishment of biotope-restricted species, to the benefit of more dry-land opportunistic species. Wetland research should give special attention to rare, localized and stenotopic species and their biotope requirements.

6.6 URBANIZATION AND INSECT CONSERVATION

6.6.1 Impact on species richness

Urbanization is a massive unplanned experiment, characterized by high human densities (>620 individuals per km^2) and an urban–rural ecological gradient (McDonnell and Pickett, 1990). The urban setting need not necessarily be depauperate of insect fauna, especially where parks and gardens dot the city. Bradley and Mere (1966) collected 367 species of Lepidoptera alone from Buckingham Palace gardens over 2 years. The faunal richness is thought to be a consequence of the plant and biotope diversity (Davis, 1978). Allen (1951–64) collected 735 species of Coleoptera in a garden in south east London over 37 years, while Owen and Owen (1975) collected 455 species of Ichneumonidae in 2 years from a suburban garden in Leicester, England.

Despite this apparently high species richness, there is some definite contrary quantitative evidence that urbanization does lower richness. Taylor, French and Woiwood (1978) found that moth faunas in urban and suburban areas are impoverished compared with woodland, while agricultural biotopes are intermediate. And rare Tettigoniidae in southern France find sanctuary in undeveloped plots and ancient farm fragments within the expanding suburbia of Montpellier, southern France (Samways, 1989a).

Vegetation and human habitation can play a role in maintaining insect diversity, but the value of this contribution is proportional to the total resources and conditions available for all aspects of insect life. Like agriculture, urbanization fragments the landscape. However, with the exception of some river banks, urbanization often has a greater fragmentary effect than agriculture. Temperature, light, humidity and wind-exposure differences, as well as high levels of local vehicle pollutants, are often quite different in urban and rural areas. This is favourable to edge rather than interior species, and those with greatest physiological tolerances.

Preserving land in the face of an expanding city has to take into account political and economic considerations as well as ecological ones. Clearly, land for urban development has high economic value, with important sociopolitical implications, which the insect conservation biologist must take into account.

On the ecological side, Murphy (1988) points out that the greatest loss of extinction-prone species has usually occurred in biotope remnants with the longest histories of urban settlement. Also, he points out that the area to be preserved to maintain diversity will depend on the type of ecosystem. Oak woodland preserves near San Francisco are likely to require more area to protect their complement of biological diversity than will the native grassland patches in the same geographical area.

Nevertheless, urban commons, parks, leisure gardens, private gardens, roadsides and cemeteries do provide refugia (Gilbert, 1989), and ecologists and landscape architects (see Bradshaw, Goode and Thorp, 1986) have much bridging to do, not just for man's amenity and for the common, eurytopic species, but also for the rarer more biotope-sensitive ones. This is especially the case in the warmer latitudes. Clark and Samways (1994) found 821 insect species in just one week's intensive and comprehensive sampling in two tennis court-sized areas in South Africa, one horticulturally landscaped with indigenous plants and the other left minimally managed. The majority of the species have yet to be taxonomically described. In all replicates, over half the species were rare, being singleton occurrences.

Generally, when new species appear in urban areas they are invasive, scavenging species, highly tolerant of a wide range of environmental conditions. Some eke out a living in the most unlikely and even precarious situations. The cockroach, *Periplaneta americana,* manages to survive quite successfully inside television sets in high-rise buildings in the centre of cities, generally avoiding high-voltage electrocution. Rare species do occasionally appear after accidental introduction, where the receiving environment is suitable. Regent's Park, London, has a colony of the rare (in England) Roesel's bush cricket (*Metrioptera roeselii*) inhabiting coarse grassland, probably the result of accidental introduction of eggs brought in with soil used for filling tree pits (Widgery, 1978).

Clearly, the above example and the unexploited opportunities for ecology along the urban–rural gradients (McDonnell and Pickett, 1990) point to there being much more to explore in insect conservation biology by ecological landscaping in municipal areas as well as private gardens. Although translocation and reintroductions have technical and ethical problems, depending on the species and distances involved, with the increased labour costs for manicure gardening, and the increasing awareness among the younger generation of green issues, there is considerable opportunity for research. In turn this would lead to practical recommendations for the home garden owner and municipal authorities to promote survival of some of the rarer as well as the cosmopolitan species.

There is also the interesting situation of urban activities actually increasing the abundance of species to pest levels. Prior to man's habitation, some of these species must have been much rarer. Examples include the *Dermestes* beetles, so familiar as destroyers of natural fabrics, dried animal skins and even pinned insects.

6.6.2 Genetic aspects

Earlier industrialization and urbanization in England led to an increase in melanic polymorphic forms of certain insects such as the peppered moth

(*Biston betularia*) and the two-spot ladybird (*Adalia bipunctata*) in response to increased pollutants in the air. The air pollution darkens tree trunks and is toxic to lichens, giving selective advantage to the dark *carbonaria* form of the peppered moth (Kettlewell, 1973). With cleaning of the air in recent years

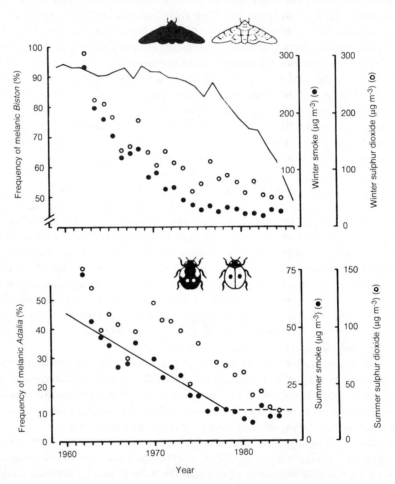

Figure 6.13 Examples of adaptive responses to environmental change involving genetic polymorphisms. Solid lines show recent declines in the frequency of melanic forms of the peppered moth (*Biston betularia*) near Liverpool, England, and of the two-spot ladybird beetle *Adalia bipunctata* in Birmingham. The declines are associated with falls in air pollution as shown by records from nearby monitoring stations. Changes in crypsis are involved in the case of the bark-resting moth, while the ladybirds are probably influenced by changes in the screening effect of smoke on insolation. (From Brakefield, 1987, 1991.)

the melanics are no longer at such a selective advantage and, in the case of the two-spot ladybird, the selective disadvantage to melanics is about 10%, illustrating how environmental change can result in intense natural selection and rapid evolutionary responses (Figure 6.13). Brakefield (1991) has also pointed out that the short generation time of many insects tends to lead to closer tracking of rapid environmental changes, at least for these temperate insects. Melanic polymorphisms also illustrate how the spatial scale of environmental change can interact with rates of migration and population structure to influence the pattern of adaptation. The peppered moth, which is highly vagile and has fairly uniform population densities, shows smooth transitions in melanic frequency between industrial and rural areas. For other species this may not be the case. The melanic gene apparently is not only dependent on selective visual predation by birds, but also has other, pleiotropic, effects on the phenotype, such as development rate and tolerance to pathogens (Brakefield, 1987, 1991).

6.7 SUMMARY

The landscape is periodically disturbed by natural events from localized tree-fall to the widespread effect of volcanic eruptions. These disturbances are discrete events that disrupt ecosystem, community or population structure, changing the physical and biotic environment. Many species of plants and associated insects are adapted to survive and thrive on such disturbances. With increasingly intense, frequent and widespread man-induced disturbances, many of the species that depend on late botanical succession (interior species) and are biotope sensitive are now suffering the consequences of the massive changes being wrought across the earth.

When populations are small and isolated, not only are they more subject to chance adverse environmental impacts but genetically they are subject to a 'bottleneck'. If these bottlenecks extend over many generations there is a detrimental loss of genes that potentially reduces the possibility of the population adapting to future environmental perturbations.

Although in the tropics and subtropics maintenance of large tracts of 'pristine' landscape is important, fragmentation has meant that many other areas have to be managed to simulate otherwise natural disturbances. These include grazing and burning.

Many insect populations show remarkably quick recovery from naturally adverse conditions such as exceptional rain storms and typhoons. It is when conditions are severe and/or prolonged, and when they are frequent, that populations begin to fragment.

For insects, different types of disturbances often have a synergistic adverse compound effect. Also, insects themselves can cause vegetation mortality and predispose the forest to other disturbances, such as fire. Insects may also interact with other agents, such as plant pathogens, which change the vegetation and, in turn, the insect community. The important point is that succession is a highly important ingredient in maintenance of insect diversity in natural ecosystems. Insects and plants, as well as other factors, are co-determinants of the successional pattern. Some of these interactions are extremely resilient, while others are not. Managing interactions intact does not guarantee survival. The county of Suffolk, England, has lost nearly half of its butterfly species, although none of their food plants has disappeared.

Both real and theoretical disturbances indicate that frequency of disturbance, particularly if of high magnitude, is especially disturbing to ecosystems. High-energy-input mechanized agriculture falls into this category, causing enormous landscape degradation and reducing the chances of recovery. Sustainability of resources is essential for future human and insect survival, fossil energy inputs are reduced, and beneficial aspects of insect biology and behaviour are encouraged. This can be achieved through modification of the plantscape and landscape, by the creation of conservation corridors and patches, which are areas of natural vegetation between crop patches. Polycultures of crops also increase natural enemy activity and reduce the need for insecticides. These agroecological techniques imply maintenance of insect diversity (and that of mites) to enhance more constant, predictable yields.

Not all aspects of agriculture are harmful, with some woodland and hedgerow management practices enhancing some insect species populations. Regularized grazing, and the creation of small farm dams in arid areas, can be beneficial.

Urbanization is generally detrimental to all but a few species of insects, principally scavengers. Such urbanization has even caused genetic changes in some species, both in response to increased industrial pollution, and then back again as the pollution level has dropped.

−7

Individual insect species and their conservation

Verily it has seemed, looking back on the past, as if
our insect hunters of the future were doomed to a
sport composed of an influx of Colorado Beetles,
White Butterflies, Hessian Flies and Woody Oak-galls;
and that the flowery wood clearings and purple heaths
of our forefathers, with their basking and fluttering
fauna, were all to be heartlessly swept away in the
present era of steam and telegraphy. It is quite certain
that our butterflies, especially the Orange Fritillaries,
are fast disappearing.

A.H. Swinton (c.1880)

There can be no finality in the scientific study of
endangered and extinct species...

HRH The Duke of Edinburgh

7.1 RARITY

7.1.1 Rarity, and its categorization

Insect species differ considerably in the constancy of their population levels. Some are relatively constant from one generation or year to the next, others are frequently varying, while others remain at fairly consistently low levels only to outbreak very occasionally (Whittaker, 1975) (Figure 7.1). We tend to think of conservation of species that are consistently rare as those in prime

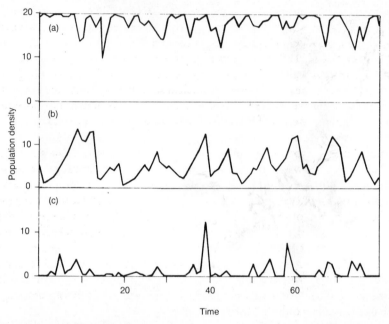

Figure 7.1 Models of fluctuation in hypothetical populations. (a) Population fluctuates near its carrying capacity ($K = 20$). Here the population is affected by environmental fluctuation within a predominantly favourable range of values that permits the population to occupy most of the potentially suitable microenvironments. This could be an insect population in a tropical forest. (b) This population fluctuates widely at a range of densities between K and L (lower limit), with occasional local extinction and reinvasion. Here the population is affected by environmental fluctuation causing population increase and decrease in the middle part of the range of microenvironments available to it. (c) The population fluctuates near the lower limit ($L = O$) and is rare much of the time, but is subject to occasional outbreaks. In this case, the population normally exists at low densities and is affected by environmental fluctuation acting on individuals in the most favourable part of the microenvironmental gradient. (Reprinted with the permission of Macmillan Publishing Company from *Communities and Ecosystems* by Robert H. Whittaker. Copyright 1975 by Robert H. Whitaker.)

need of protection. However, rarity has different complexions, and insects with different life-history styles may need protection. Additionally, in conservation, rarity has a sociopolitical element in that a species may be listed as threatened in one country yet be quite common in another. Selecting sites for particular rare species is one approach towards site-selection for conservation of communities.

Rarity for conservation evaluation has several facets and, as Usher (1986b) has pointed out, rarity may not be particularly important scientifically but is of critical practical importance. Rare species cause a strong emotive reaction in many quarters, and provide strong political weight. But when selecting a site for conservation, some species may be rare there and others rare elsewhere even within a relatively small geographical area.

Schoener (1987, 1990) recognized 'suffusive rarity' (i.e. species that are rare in particular localities throughout their geographical ranges) and 'diffusive rarity' (i.e. species that are common at some localities but rare at others). For insects, suffusive rarity is appropriate where particular biotopes are widespread and where the niches of the insect species are narrow, and possibly also where interspecific competition keeps the populations low.

Gaston and Lawton (1990) have pointed out that for insects there is either no relationship between local abundance and geographic range or, more commonly, a positive relationship. In other words, a locally common species generally has a wide geographic range. Species that exploit a wide range of resources become both widespread and locally abundant, while species that occur in an original experimental biotope (reference biotope)–that is not sufficiently different from other biotopes in the region–may have a small or a large geographic range. Another category is where the reference biotope is distinctly different from other biotopes in the geographic region. This results in a negative correlation between local abundance and geographic range. Some cave insects would fall into this category.

Rarity is therefore closely tied with the biotope-tolerant (eurytopic) to biotope-specific (stenotopic) spectrum. A strong positive correlation between abundance and biotope tolerance has been shown for ants (Samways, 1983a) (Figure 7.2). There is the self-evident fact that species with wider biotope tolerances are more able to be more abundant overall. With these ants, the eurytopic species have the widest geographic ranges (Samways, Nel and Prins, 1982). However, at the behavioural level, rarity, at least in ant assemblages, is dictated by competition which is asymmetrical to completely amensalistic (i.e. total outcompeting of one species by another) (Samways, 1983c). This may lead to local mosaics of dominant species (Leston, 1973) or, on a larger scale, geographical replacement (Samways, Nel and Prins, 1982; Samways 1983c). In terms of insect conservation biology, however, most of the rare ants remain rare throughout their ranges.

For comparative purposes, it is valuable for insect conservation biology

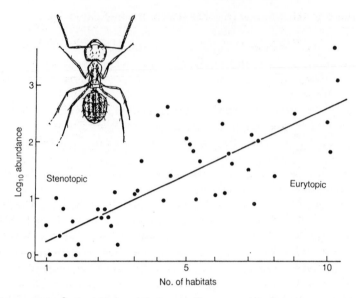

Figure 7.2 Log-abundance of 44 South African ant species relative to the number of habitat structures in which they occur ($R = 0.594$, P <0.05; $F = 71.2$, P <0.0001). (From Samways, 1983a.)

also to look at what the botanists perceive as rarity values, as insects are small animals with generally small home ranges and mostly directly or indirectly dependent on plants.

Rabinowitz, Cairns and Dillin (1986) have examined types of rarity, and in what important ways rare species differ from one another. They first distinguish three traits characteristic of all species.

1. **Geographical range.** Whether a species occurs over a broad area or whether it is endemic to a particular small area.
2. **Biotope specificity.** The degree to which a species occurs in a variety of biotopes or is restricted to one or a few specialized sites. (Note that Rabinowitz, Cairns and Dillin (1986) used the term 'habitat' for 'biotope'.)
3. **Local population size.** Whether a species occurs in large populations somewhere within its range or has small populations whenever it is found – not unlike diffusive and suffusive rarity respectively.

Based on these three traits Rabinowitz, Cairns and Dillin (1986) have drawn up seven forms of rarity:

Insect species for which we have sufficient records fall into one of these categories. A plant species may be recategorized either because it changes itself, e.g. through genetic changes, or because it is changed by an outside

Table 7.1 Rabinowitz et al.'s (1986) seven forms of rarity

	Geographical range	Biotope specificity	Local population size
1.	Wide	Broad	Somewhere large
2.	Wide	Broad	Everywhere small
3.	Wide	Restricted	Somewhere large
4.	Narrow	Broad	Somewhere large
5.	Narrow	Broad	Everywhere small
6.	Narrow	Restricted	Somewhere large
7.	Narrow	Restricted	Everywhere small

agency, particularly by man breeding it, influencing its biotopes or translocating it accidentally or deliberately. Insects will change categories even more readily than plants because of their relative population variabilities, particularly in response to landscape changes. Others will migrate in response to landscape modification, or in response to other biotic or abiotic variations, particularly pathogens and weather.

Rabinowitz, Cairns and Dillin's (1986) classification is a useful conceptual framework for insect conservation biology. It helps to view changes in insect species status as it moves from one category to another. Such changes should be seen against the background of the preceding distribution pattern. However, the dynamic nature of distributional abundance, whether in the shorter term under the influence of man-influenced landscapes, or in the longer term under the influence of climatic or geological changes, make fixing of a species into a rarity category artificial in the truly biological sense.

7.1.2 Rarity and preservation of genes

A species may be rare simply because it is restricted by a distinct biotope, such as a pond or a small island. Over time, with constancy of environmental conditions and with little or no gene flow, this restriction tends to endemism. Such species, being adapted to evenness of surrounding conditions and being genetically remote, are particularly threatened. They are among the species physiologically least able to adjust to man-induced changes, whether local landscape modification or global climatic change. Yet they are also among the most intrinsically valuable species because of the uniqueness of much of their active genome.

In geographical areas with high levels of plant species endemism and diversity, there are often similar trends in insect herbivores. The high plant diversity does not necessarily directly generate high insect diversity. A more accurate interpretation, although closely associated, is that there are processes common to both taxa. Such processes include allopatric speciation through area isolation effects caused by the plants themselves or by topographic variations against a long historical background of climatic constancy (section 1.5). These allopatric effects can be strong in the tropical forest and the Cape Floral Kingdom, to the extent of causing species patchiness to the same extent as does a small island or a cave.

Management response to rarity in the conservation context has different values depending on the sociopolitical setting. Considerable research and concern has been expressed at the loss of the large blue butterfly (*Maculinea arion*) from Britain, with the species having been reintroduced from Swedish stock (Thomas, 1983b). While admirable, such activities illustrate the differences between the species-poor, economically wealthy north-temperate areas and the developing world, where the genetic loss is enormous each year. If, say, a species in the equatorial area carries 10 000 genes, and conservatively 1000 species per year are being lost, the genetic loss runs into ten million alleles per year. Caution against exaggerated claims of loss is needed at the level of the genes, as many insects possibly have at least 80% of their genomes in common. Many of these shared genes would be inactive, while others would be controllers of biochemical and physiological processes fundamental to insect function.

The immediate, realistic response towards stopping this loss of genes is to document the distribution and rarity of as many species as possible, and to preserve the maximum number of relatively undisturbed landscapes. This task is not entirely feasible in view of the rate of global loss of species-rich biotopes. Twenty thousand species are being lost per year. This could be overly pessimistic. Brown (1991) points out that, despite over 95% of the Brazilian Atlantic forests having disappeared or having been strongly disturbed, not a single butterfly or dragonfly can be definitely declared extinct. To this should be added the rider that many tropical species are extremely rare, making it difficult to know whether a species is really extinct or simply has not been seen for many years.

Although the main aim is to preserve as many pristine biotopes as possible, with global atmospheric impacts and land fragmentation species ranges will inevitably fragment and shrink. This means conserving areas where rare species are common in the case of diffusive rarity, and to predict where climatic change is likely to be minimal in the case of suffusive rarity.

7.2 THE TAXONOMIC IMPEDIMENT

The lack of scientific names and the difficulty in identifying species in an ecological study is the taxonomic impediment (New, 1984). At least 80% of all vertebrate species have been described, compared with only 5–10% of the insects. Wheeler (1990) has produced a highly visual 'speciescape' ('species-scape' is a better alternative spelling) in which described numbers of species

Figure 7.3 The 'species-scape' of Wheeler (1990) in which the size of the organism represents the size of the described species in the overall taxonomic group (principally at class level). The beetle, for example, represents the insects; the elephant the mammals; the mite the non-insect arthropods; the pine trees the multicellular plants, etc. NB As so many insects are yet to be described, compared with the other groups, in real terms the beetle in this illustration should be much larger.

of insects are illustrated in proportion to the described numbers of species of other groups (Figure 7.3). This succinctly illustrates how important insects are on the landscape and in ecosystems, and has an even greater impact when one realizes how much larger 'the insect' in Figure 7.3 would be if all the insects were scientifically described.

Such a species-scape is a valuable conceptual image for overcoming the lack of understanding of the major significance of insects in ecosystem functioning.

The 22 400 British species that have been described probably represent 99% of the resident extant taxa. In the tropics, Stork's (1988) estimate is quite the converse, less than 1% of all species have been described. The subtropical countries are intermediate, with the 80 000 insect species described in southern Africa (Prinsloo, 1989) representing possibly a quarter of the total. Even well-collected and studied groups such as the Lepidoptera are only about 80% described. In Australia, 48 600 species have been described, of the estimated total of at least 125 000 species (Greenslade and New, 1991).

The significance of taxonomy cannot be overemphasized and, with only about 1500 professional systematists in the world competent to deal with the millions of tropical species (E. O. Wilson, 1988), the likelihood of ever naming many of the insects that are co-habitants with us on earth is slim.

This taxonomic impediment is one of the greatest hurdles facing insect conservationists. It must be addressed urgently. In the future many newly described species are likely to be already extinct, if not of indefinite rarity status.

Wheeler (1990) has emphasized that this impediment may mean that 50% of insects will become extinct in the next 30 years, which is an hourly loss of 19 species, a daily loss of 456, and an annual loss of 167 000. These are broad estimates illustrating the urgency for much systematic work. This 'expertise crisis', as Wheeler (1990) calls it, is truly highlighted when the number of taxonomists is compared to the number of species within their speciality. For entomologists this is one taxonomist for every 425 species. But the reality is that there are far more insect species than originally thought. If the total estimate of 10 million insect species is considered, the ratio of taxonomist to insect species becomes 1:5669. If the higher, possibly overliberal, figure of 50 million extant species is taken, the ratio then rises to an enormous 1:28 345. This is quite a responsibility for our present generation.

7.3 OFFICIAL CATEGORIES OF THREAT

Worldwide it has been generally standard practice to categorize threatened species and phenomena according to the following IUCN criteria (Wells, Pyle and Collins,1983):

Extinct (Ex). Species not definitely located in the wild during the past 50 years.

Endangered (E). Taxa in danger of extinction and whose survival is unlikely if the causal factors continue operating. Included are taxa whose numbers have been reduced to a critical level or whose 'habitats' have been so drastically reduced or modified that they are in immediate danger of extinction. Also included are taxa that are possibly already extinct but have nevertheless definitely been seen in the wild in the past 50 years.

Vulnerable (V). Taxa believed likely to move into the 'Endangered' category in the near future if causal detrimental factors continue operating. Included are taxa of which most or all populations are decreasing or being lost to disturbance and fragmentation; and taxa whose biogeographical range is decreasing through loss of peripheral populations.

Rare (R). Taxa with small or limited geographical range populations that are not at present 'Endangered' or 'Vulnerable', but are at risk.

Indeterminate (I). Taxa known to be 'Endangered', 'Vulnerable' or 'Rare' but where there is not enough information to categorize.

Out of Danger (O). Taxa formerly in one of the above categories, but which are now secure because conservation measures have removed the threat, notwithstanding global climatic changes in the future. No insects are included in this category in the *IUCN Invertebrate Red Data Book* (S.M. Wells, Pyle and Collins, 1983).

Insufficiently Known (K). Taxa that are suspected but not definitely known to belong to any of the above categories through lack of information. The vast majority of insects, by default, fall into this category.

Commercially Threatened (CT). Taxa not currently threatened with extinction, but most or all whose populations are threatened as a sustainable commercial resource. This category is not used if any of the traditional categories apply. In practice, it has only been used for marine species of commercial importance being overfished in several parts of their ranges. Some rare butterfly specimens have been advertised for as much as US $7000 each, and South African *Colophon* stag beetles for US $ 11000. Also Taiwan has annual sales of between 15 and 500 million butterflies to be laminated for ornaments and decorations, while 50 million butterflies per year being killed in Brazil (New, 1987), it is possible that this category will apply to some insect species in the future.

Threatened Community (TC). A group of ecologically linked taxa occurring in a defined area, which are all under the same threat and require similar conservation measures. Several TCs are listed by Wells, Pyle and Collins (1983), including the Usambara Mountains of Tanzania (with their rich plant and animal endemic fauna), the rain forests of Gunung Mulu, Sarawak, Malaysia (which includes examples of almost all Sarawak's inland forest types), the Puerto Rican cave Cueva los Chorros (with its rich invertebrate fauna, which is vulnerable to disturbance because of the cave's small size), the mines of the Sumava Mountains (former Czechoslovakia) (with their scattered montane peat bogs being refugia for relict mire boreo-alpine and subarctic Lepidoptera and other insects) and some other communities.

Threatened Phenomenon (TP). Aggregates or populations of organisms that together constitute major biological phenomena, endangered as phenomena but not as taxa. The best-known are the Californian and Mexican winter roosts of the migratory monarch butterfly (*Danaus plexippus*). Also listed by Wells, Pyle and Collins (1983) is the Japanese wood ant (*Formica yessensis*) supercolony, composed of a huge complex of 45 000 closely associated *F. yessensis* nests on the Ishikari Coast, Hokkaido, threatened by building developments.

These categories are widely used and have become the general standard categorization of threatened species worldwide. In the USA however, a different terminology is used and framed in the Endangered Species Act of 1973, which is a legislative effort intended to preserve ecosystems upon which endangered species depend (Pyle, Bentzien and Opler,1981). Any insect, with exception of pest species, can be afforded the Act's protection. An Endangered species in the USA is a species in danger of extinction throughout all or a significant portion of its range, while a Threatened species is one likely to become Endangered in the foreseeable future (Opler, 1991).

The general umbrella term 'threatened' with a small 't' is useful and widely used for species tending towards dangerous population reduction. For the majority of insects, we have no scientific names, let alone knowledge of their status. The species officially categorized are mostly large and conspicuous, from well-known areas (as are most of the species listed in Wells, Pyle and Collins, 1983), or are from thoroughly searched countries or areas (as the compilation for the British Isles by Shirt, 1987).

The IUCN categories are not necessarily definitive. There is some criticism of the categorizations, which according to Mace and Lande (1991) are subjective, and with inherent judgement risks. In turn this could lead to wrong planning decisions.

Recent developments in the study of population viability/vulnerability have resulted in techniques that can be more helpful in assessing extinction risks. Mace and Lande (1991) redefine categories in terms of the probability of extinction within a time period, based on the theory of extinction times for single populations and on meaningful time scales for conservation action. Three categories, in addition to Extinct, are proposed:

Critical. Fifty per cent probability of extinction within 5 years or 2 generations whichever is longer.
Endangered. Twenty per cent probability of extinction within 20 years or 10 generations, whichever is longer.
Vulnerable. Ten per cent probability of extinction within 100 years.

These are only suggestions at this stage, and definitely have merit with regard to vertebrate populations. For insects different criteria may be needed on the basis of their different overall biology. Insects often have high population growth rates, short generation times, great fluctuations in population abundance and distinct biotope preferences. With these points in mind it will be more important to incorporate metapopulation characteristics such as subpopulation persistence times, colonization rates and the distribution and persistence of suitable biotopes into the assessment (Murphy, Freas and Weiss, 1990).

With the taxonomic impediment being so great and time so short, there is a real need to rapidly identify areas with high numbers of endemic species, areas that are home to a large number of typical species, and also to recognize areas that are dynamos for species diversification now and in the future. There is then some merit in extending the term 'Threatened Community' to many more individually identified sites until such time that a more thorough inventory of insect species can be made. Even better in terms of terminology, at least for terrestrial ecosystems, would be 'Threatened Landscape'. This implies land management as well as straight preservation. It also encompasses anthropogenic landscapes and landscape elements of conservation interest, such as coppiced woods, ancient hedgerows and ancient human settlements, in addition to undisturbed tracts of tropical forest, to realistically conserve all species alongside man's activities.

7.4 THE *RED LIST* AND *RED DATA BOOKS*

7.4.1 The *Red List*

The IUCN Red List of Threatened Animals (IUCN, 1990) is compiled by the World Conservation Monitoring Centre in Cambridge, UK. It is an authoritative list of those taxa known by IUCN to be threatened. It complements the IUCN *Red Data Books* and the IUCN Action Plans, both of which contain more detailed information on the conservation status of species. The list is published periodically, listing threatened taxa and their categorization, and a brief description of their distribution.

The information is compiled by many concerned biologists through published and authentic unpublished data. Importantly, much of the information is collated by the Specialist Groups of the IUCN's Species Survival Commission. Naturally, there are biases towards certain groups, especially the butterflies and dragonflies, which have a high profile through their size, colour and relative ease of identification. The *Red List* also reflects the relative degree of active participation by workers devoted to certain groups. This bias is a natural manifestation of our lack of overall knowledge. Nevertheless, the *Red List* is an important foundation which will continue to be updated and refined as further reliable data from active participants become available. Also, the taxa listed are only a fraction of those that should be given mention, and therefore the *Red List* will inevitably increase in size not only as more information is obtained, but also because of increasing threats to more species.

The *Red List* is also useful, as is the *IUCN Invertebrate Red Data Book,* for highlighting threatened insects that are in danger because they are ectoparasites on threatened vertebrates (potentially even on other threatened invertebrates). Such an example is the pygmy hog sucking louse (*Haematopinus oliveri*), which occurs only on the 'Endangered' pygmy hog (*Sus salvanius*) from north-west Assam (Wells, Pyle and Collins, 1983). The *Red List* is unquestionably an important barometer of the biodiversity crisis by listing threatened species and their status. It is an important document, not simply advertising gloom, but pointing to areas of further study and towards management, particularly of taxonomically remote and interesting species.

7.4.2 *Red Data Books*

The *Red Data Book* idea originated with Sir Peter Scott in the mid-1960s, with the publication of *The Red Book: Wildlife in Danger* (Fisher, Simon and Vincent, 1969). The foreword by H.J. Coolridge and P. Scott was '...written neither in optimism nor in pessimism, but simply objectively, as part of the duty to focus world attention on wildlife in crisis....'. The volume did not cover insects, but since then *Red Data Books* (RDBs) have blossomed, with volumes covering invertebrates in general (S.M. Wells, Pyle and Collins, 1983), world status of a specific insect group, the swallowtail butterflies (Collins and Morris, 1985), national status of individual insect groups (e.g. Henning and Henning, 1985) and national status of all threatened insects (Shirt, 1987).

None of these volumes claims to be definitive, particularly with such a cryptic, little-known and diverse group as the insects, often with their considerable sensitivity to biotope change. These RDBs are workbooks to stimulate further confirmation of the status of individual species and as management handbooks. Naturally, they are backed up by other publications, such as local reports (e.g. Collier, 1986) and original research papers, all taking a step towards preservation of the insect fauna.

Ferrar (1989) has provided a valuable critique of RDBs. He points out that, as well as identifying rare and declining species and the extent and nature of their decline, and the setting priorities for conservation, monitoring and research. RDBs also draw public attention to conservation issues by highlighting the plight of individual species. This political or public awareness role of RDBs achieves its ends by means of environmental education, political pressure groups and fund raising, i.e. they have more than just a passive role.

Ferrar (1989) also points out a paradox with RDBs. Listed invertebrate species often have highly restricted distributions down even to a few hectares, the exact location of which is rarely mentioned in order to keep the sites confidential. This may be necessary to avoid such negative influence such as a landowner ploughing up a site of specific interest to forestall a protection order or to inflate the selling price of the site. This is, of course, not always the case, and positive responses often arise, with landowners protecting or managing a specific site.

The information contained in RDBs is a guide. A critical eye is needed to determine whether a national RDB lists a species simply because that country is on the far edge of the range of the species. This is particularly the case with many British species, which although listed as threatened in one or other category may in fact be quite common in continental Europe.

This is where the concept of endemism becomes important. National RDBs do, in some facets, have a political bias rather than a biological one. Listing of a taxonomically remote rare endemic is, in genetic terms, far more pertinent than, say, a national rarity on the edge of its range. Listing of introduced species also occurs, when they are rare. For example, the scaly cricket (*Mogoplistes squamiger*) is listed as 'Endangered' in Britain, and is restricted to a single locality (Haes, 1987). It could be a relict species hanging on in Britain, or a species of the Mediterranean littoral and Madeira. But more likely the British population was introduced via the nearby Portland Naval Base during the Second World War (Haes, 1987).

The level of threat as well as the number of listed species occurring in a restricted geographical area are among the valuable pieces of information supplied by RDBs. They do help identify important taxonomic groups and geographical areas, although for insects many of the assigned categories are informed guesses. Nevertheless, such RDBs on insects are an extremely valuable start. For example, the high number of listed endemic threatened species of lycaenid butterflies in the south-western Cape (Henning and Henning, 1989) immediately draws attention to the vast genetic variety at stake in a highly restricted geographical area.

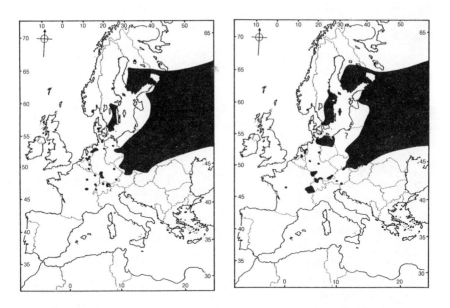

Figure 7.4 Blocked-in range distribution map of two species of dragonflies [*Leucorrhinia caudalis* (left) and *L. albifrons* (right)] with very similar ranges. Even for such well-known insects as the European Anisoptera, there are still gaps in our overall distributional knowledge, as in the case of *L. albifrons* on the eastern shores of the Baltic. (From Askew, 1988.)

7.5 DISTRIBUTIONAL RECORDS

7.5.1 Coarse-scale maps

Determining the distribution of species is essential for making decisions on their conservation management. The spatial scale of measurement has a major impact upon perceptions of rarity, threat and stability of populations. The generalized distributional maps, familiar in bird identification handbooks, are of limited value for insects, with their smaller body size, small home ranges and specific biotope requirements. Nevertheless, some good examples are available which give a useful overall picture of the geographical range of a species (e.g. Askew, 1988, for dragonflies of Europe) (Figure 7.4).

One of the values of such blocked-in range maps is that they indicate whether a species has a distribution which extends into a particular country. Such an insect may then be included in the *Red Data Book* for that country. Overall, such a species may not be rare. It may be at the edge of its range, where environmental pressures may well be greatest, particularly in adverse years. Over periods of several decades, and also with major changes in global climate imminent, such coarse-scale distributional maps are useful in giving an overall view of shrinkage points and where important preservation areas might be.

There is a problem with such maps in that the data on which they are based may be weak or incorrect. The information may be old, based on casual observations, or involve a generalized semi-guessing of blocking in. For most insects, even in well-explored areas, there are no precise data on distribution, and generalized coarse-scale maps really only indicate where a species might be found.

Even where there are very large-scale maps, e.g. for the tettigoniid wart-biter *Decticus verrucivorus* (Samways and Harz, 1982) (Figure 7.5), it is only where there are fine-scale studies of the particular species in a particular area that the overall threat to the species can be appreciated. *D. verrucivorus* is an abundant insect in the Cevennes Mountains of southern France, yet categorized as 'Vulnerable' with a highly localized distribution and precarious status in Britain (Cherrill and Brown, 1990a). *D. verrucivorus* presents some interesting perspectives relative to distribution and insect conservation. In Britain, as in the Cevennes Mountains for example, the subspecies is *D.v. verrucivorus*. It is one of Britain's most threatened orthopterans and one of the largest of the country's insects. Once common across southern England, it now only occurs at a few guarded sites in Wiltshire, Dorset and Kent (Marshall and Haes, 1988). It is also being captively reared as a joint venture between the London Zoo's invertebrate conservation centre and English Nature. Other subspecies of *D. verrucivorus* include subsp. *monspeliensis* from Mediterranean France (zone 17 in Figure 7.5). In the past, *monspeliensis* has reached such high proportions that it has been a pest of grain crops, yet today it appears to have completely disappeared (Samways, 1989a). Elsewhere in the range of the species, little is known of the behaviour and population dynamics, but there are several distinctive morphological subspecies with highly restricted ranges. In contrast to the high conservation profile of the species in Britain, which is at the western edge of its range, elsewhere at the edge of its range, particularly in the east (Siberia) and south-east (Mongolia), almost no information is available on its status.

Although coarse-scale maps illustrate overall distribution, and may indicate fragmented distributions and also where further searches should be conducted, they provide little locality or biotope data on population dynamics or on localized conservation status of the species concerned.

7.5.2 Climatic mapping

Climate diagrams of Walter (1985) and Walter and Leith (1960–67) combine temperature and rainfall patterns into a single diagram and, although not the only means of representing climatic types, they have nevertheless proved to be valuable in both the biogeographical (Cox and Moore, 1985) and conservation contexts (Kingdon, 1990) (Figure 7.6). They also have been applied to describing the range expansion of a ladybird, *Chilocorus nigritus* (Samways, 1989d). It is an economically important predator of many scale insects. The beetle is indigenous to the Indian subcontinent and Indo-China. In the late 1930s the beetle was introduced as a biocontrol agent into Mauritius and the Seychelles. Since then it has spread to neighbouring islands and mainland Africa. Similarly, there has been an eastwards expansion into several of the Pacific Islands. Recently, the beetle has also appeared in north-east Brazil and West Africa. The important point here is that, although the species is a good invader, it has specific climatic tolerances. It is restricted to Walter's zonobiome II in southern Africa. Research has shown that it principally only establishes in this zonobiome, and zonobiome I on tropical islands.

From an insect conservation biology viewpoint, climate diagrams could prove useful, as they have been for this biocontrol agent, in determining climatic range cores and edges of particular species. The map versions of the climate diagrams also lend themselves to being an overlay in geographic information systems and so are useful in conservation management.

Other climate mapping tools available to conservation biologists include the BIOCLIM system (Busby, 1991). This can predict distribution of individual entities such as species or vegetation types. BIOCLIM is based on continuous mathematical surfaces fitted to measured meteorological data, and can be used to generate estimates of monthly mean minimum and maximum temperatures and precipitation for any point on or near mainland

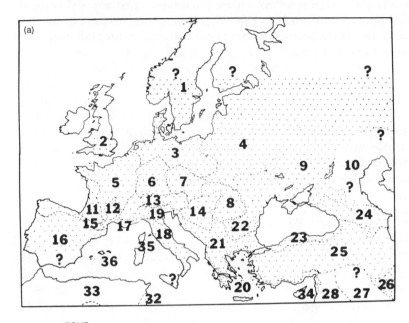

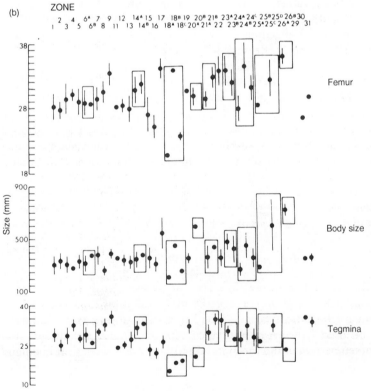

Australia and Tasmania, from inputs of latitude, longitude and elevation. CLIMEX (Sutherst, 1991) is another generic model which can predict a species response to climate. Our laboratory is currently using this programme with outstandingly reliable results.

7.5.3 Migrant species

Many migrant insect species reach high population levels and may become agricultural pests. Migrant insects are often spatiotemporally dynamic, crash-boom population species, as well as being invasive. They are therefore not generally of conservation concern, except in the special cases where the species has special roosts which are threatened, e.g. the monarch butterfly (section 4.6). Other insect groups congregate in specific areas, coccinellid beetles being a notable example (Hodek, 1973). Although their roosting sites are not listed to date, they warrant investigation by insect conservationists and recommendations should be made as to their preservation. Hodek's (1973) photographs clearly show the importance of the topographic landscape (toposcape), with ladybirds congregating in and among rocky crevices on hilltops, high above agricultural land.

These agriculturally valuable species roosts also should be classed as

---

Figure 7.5 (a) Distribution of the wart-biter bush cricket, *Decticus verrucivorus* (stippled) according to the zones:
1, Scandinavia; 2, British Isles; 3, Germanic Plain;, 4, Poland/North-West Russia; 5, French Plain; 6, Upper Rhine and Danube Catchment areas; 7, Bohemian-Moravian Highlands; 8, Carpathians; 9, Ukranian Steppes; 10, Caspian Steppes; 11, Biscay Lowlands; 12, Massif Central; 13, Alps; 14, Hungarian Plain; 15, Pyrenees; 16, Iberian Peninsula; 17, French Mediterranean Plain; 18, Apennines; 19, North Italian Lowlands (Po Plain); 20, Greek and Turkish Mediterranean Coast, Aegis and Crete; 21, Dinaric, Pindo and Balkan Mountains; 22, Danube Plain; 23, Pontic Mountains; 24, Caucasus; 25, Asia Minor (Anatolia and Armenia); 26, Iranian Highlands; 27, Tigris Plain (Mesopotamia); 28, Syrian Desert; 29 (not shown, Mediterranean Near-East; 30 (not shown), West Asia (to 80°E); 31 (not shown), East Asia (from 80°E); 32, Lowland North Africa; 33, Atlas Mountains; 34, Cyprus; 35, Corsica and Sardinia; 36, Balearic Islands; 37 (not shown), Canaries, Madeira and Azores. Question marks indicate where isolated populations may occur. (b) Dimensions (mean ±1 SD) of femur length, body size (keel length × pronotum width × gena width) and tegmina length of male *Decticus verrucivorus*. Coordinates without boxes are two or three subspecies found within the one zone; 6A, *v. verrucivorus*; 6B, *v. mesoptera*; 14A, *v. verrucivorus*; 14B, *v. gracilis*; 18A, *v. aprutianus*; 18B, *v. mithati*; 18C, *v. verrucivorus*; 20A, *v. mithati*; 20B, *v. loudoni*; 21A, *v. verrucivorus*; 21B, *v. crassus*; 23A, *v. stoljarovi*; 24B, *v. gracilis*; 24C, *v. verrucivorus*; 25A, *v. stoljarovi*; 25B, *v. crassus*, 25C, *v. verrucivorus*; 25D. *v. mithati*; 26A, *v. annelisae*; 26B, *v. verrucivorus*. (From Samways and Harz, 1982.)

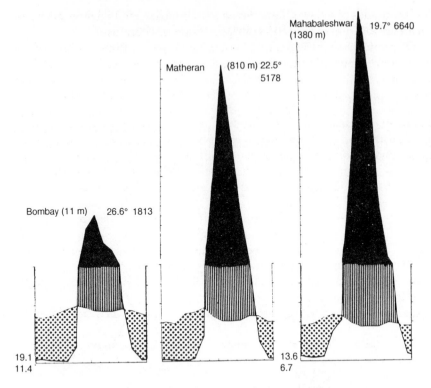

Figure 7.6. Climate diagrams illustrating the increase in quantity of precipitation at three elevations (11 m, 810 m and 1380 m above sea level) in the monsoon region of India. The central curve represents the mean monthly precipitation, from January to December, in 20-mm divisions. The more even curve in this figure is the mean monthly temperature in 10°C divisions. The shading makes for clear reading of the climate diagram. The stippled area illustrates dry winters, while the lined and blocked areas illustrate peak summer rainfall. (From Walter, 1985.)

'Threatened Phenomena' where they form major congregation sites subject to human population pressure. These ladybirds are important natural enemies of aphids and other Homoptera, and thus contribute towards sustainable agriculture in the spirit of *Caring for the Earth* (IUCN, 1991).

7.5.4 Atlassing

Atlassing or mapping of insects is far behind that of vertebrates. Various countries have developed their own systems based on various attributes, criteria and values (Spellerberg, 1981; Usher, 1986a; Noss 1987). The question of scale is important, and for insects at the biogeographical level (i.e. thousands of square kilometres, whole countries, etc.) a dot system based on actual location of collection or observation has been used.

Of particular note is the grid system employed in Britain. Maps of the country are prepared by the Biological Records Centre (BRC), Monks Wood Experimental Station. The Centre began in 1954 as the Distribution Maps Scheme of the Botanical Society of the British Isles and resulted in the publication of the *Atlas of the British Flora* in 1962. The Atlas is a compilation of distribution maps for each of the plant species using a dot per 10 km^2. In 1968, the late John Heath initiated the 'Insect Mapping Scheme' beginning with Odonata and the 'orthopteroid orders'. In later years the BRC passed

Figure 7.7 Localities from which there are records of Lepidoptera in Bavaria, Germany, regardless of time and whether the species is extinct or extant at the present time. (From Kudrna, 1986.)

on responsibility for national recording schemes to volunteers. Sections of the scheme have been very successful, much more so than some of the European schemes, which are inadequate and outdated (Kudrna, 1986) (Figure 7.7).

One of the most complete BRC programmes has been the Orthoptera Recording Scheme initiated in 1977, producing a regular newsletter and a comprehensive collection of maps (Marshall and Haes, 1988). Cards are used for recording the basic reliable data. Computer files are then compiled with information on locality name, grid reference, county, recorder, date of occurrence and biotope. The computer files are then used to compile distribution maps and to sort information under various headings for conservation, planning and research.

Records on the maps for the Orthoptera recording scheme are plotted with characteristic symbols:

O   records of 1960 or earlier
●   records of 1961 or later
■   post-1980 records for certain *Red Data Book* species
X   post-1960 introductions
?   record is probable but unconfirmed.

Figure 7.8 gives the approximate past and present localities of the Red-Listed wart-biter bush cricket in Britain.

As Marshall and Haes (1988) point out, the mapping of British orthopteroids over the last 20 years has been a relatively simple task. There are comparatively few species to consider, the majority can be recognized in the field, and the data have been gathered with enthusiasm and thoroughness. Most post-1960 records are accurate, while the pre-1960 records have caused problems of accuracy.

Such a mapping scheme is nevertheless invaluable and, in general, indicates that most species of orthopteroids, as well as other groups, e.g. Lepidoptera, in Britain have undergone range fragmentation and shrinkage with increasing rapidity this century.

The development of such maps for many countries and areas, especially the tropics, is out of the question in practical terms at the present time. But there is no reason why it should not be started in selected areas, for example in large game reserves and other protected or biologically important areas.

7.5.5 Fine-scale maps

Even a 10 km^2 does not generally specify the true fine-scale distribution of a species, which is determined rather by the distribution, shape, size and connectivity of its biotope fragments. A useful illustrative step is to use different symbols to illustrate the number of sites per 10 km^2. Oates and

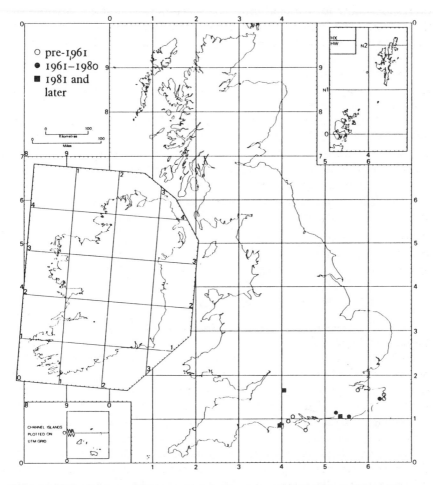

Figure 7.8 Distribution of the wart-biter bush cricket (*Decticus verrucivorus*) in Britain, based on dots representing 10-km squares. (From Marshall and Haes, 1988.)

Warren (1990) used this form of atlassing to show where the wood white butterfly (*Leptidea sinapis*) had been introduced and had established in addition to its sites of natural occurrence (Figure 7.9).

Further, it also depends on the tactics and strategies of the species themselves. In ecological time, the biotope provides the templet on which evolution forges characteristic life-history styles (Southwood, 1988). Most biotope templet proposals have two axes, one being the frequency of disturbance and the other the general level of adversity or harshness of the environment. In other words, although disturbance and fragmentation

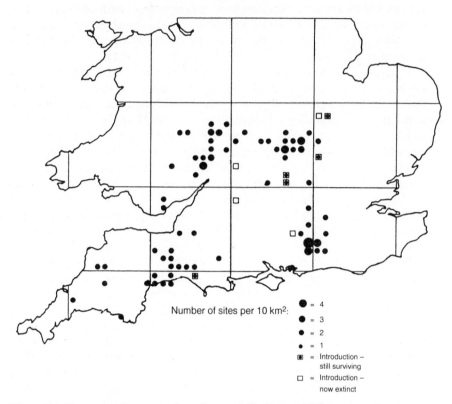

Figure 7.9 The approximate number of post 1960 sites per 10 km square in England and Wales where the wood white butterfly (*Leptidea sinapis*) survives. (From Oates and Warren, 1990. Updated from Warren,1984.)

change biotopes, different species, because of their different life-history styles and responses to biotope change, will be differentially affected. The change may be critical for only one part of the life cycle, or by preventing one necessary behavioural activity, such as nectar feeding, oviposition, mate meeting etc. A crucial break in these life activities may mean loss of the small population, and perhaps loss of the 10 km^2 dot on the distribution map.

Although the biotope may be defined in terms of two abiotic axes (frequency and level of disturbance), the scaling of these in time and space must be appropriate to the temporal (e.g. generation time) and spatial (e.g. home range) scales of the organism being considered. An additional consideration is that the trade-offs will be constrained by the available genetic variability, with the possibility of there being more than one stable strategy for a particular environment (Southwood, 1988). In insect conservation terms, this implies a detailed knowledge not only of the biology of individual species, but also of how, when and where they interact with other organisms (including intraspecifics), as well as how they are affected by abiotic variables. The biotope combined with the insect species' behavioural ecology and its life-history style make up the insect's habitat (section 4.1).

Fine-scale mapping, by walking and making counts, has been valuable in assessing the population levels of certain species of insects, principally threatened Lepidoptera in Britain (Pollard, 1982) and the USA. In practical terms, the urgency of slowing the biodiversity crisis, mapping and recognition of landscape, biotope and biotope types and the species they support is probably the most immediate realistic goal (Clark and Samways, 1992).

Time is too short and funding too scarce to support more than a handful of long-term autecological studies. The biotopes in turn need to be related to overall landscape structure and function to provide knowledge of their viability for particular species. Insect populations, even of relatively common species, do crash in response to biotic and/or abiotic factors. Dempster's (1975) studies on the common cinnabar moth (*Tyria jacobaeae*) are particularly illuminating, with heterogeneity of the moth population and of the biotope, coupled with adult mobility, preventing extinction at the sites. However, at one site the population did become extinct as a result of heavy flooding one year. This emphasizes the significance of the heterogeneous landscape and connecting corridors in preventing range shrinkage. This is especially so as some apparently mobile insect species, such as some butterflies, are not so readily invasive of new biotopes as is generally thought (J.A. Thomas, 1991). It is the harsh events, when too severe and regular, that promote local permanent extinction and hence range fragmentation, reduction and finally overall extinction.

7.5.6 Geographic information systems

Geographic information systems (GIS) are extremely valuable in determining biotopes and landscapes for conservation. GIS are a tool for analysing and updating spatial information quickly and efficiently. They use computer-based techniques for encoding, analysing and displaying multiple data layers. A good example is Yonzon, Jones and Fox's (1991) study of the red panda 'habitat' in Nepal. They digitized and overlaid information from maps of land cover and utilization, elevation and aspect favourability to give a core habitat map. Then core areas at risk were identified by overlaying a map of land cover and utilization within 500 m of core biotopes. The technique showed that 60% of the core habitat was at risk from human pressure, such as grazing and firewood collection.

Such methods could be used for distributional mapping of selected species of insect, and levels of risk to the mapped populations. But as considerable time would be spent on finding all the sites where the species occurs in the field, GIS probably have more value in determining biotopes where certain species, assemblages or communities are likely to be found. This may mean principally conserving plantscapes as an umbrella strategy for the local invertebrates. As with the red panda example, risk analysis can then be carried out to assist insect conservation management decisions (section 10.4).

7.6 SUMMARY

Insect species differ considerably in the constancy of their population levels. A species may be rare in one area but not in another. Some are always rare residents, some always common, and some mostly rare but outbreaking occasionally to high population levels. Many are also migratory. Rarity over time must therefore be related to the life-history style of the species. Conservation decisions hinge on an appreciation of the population dynamics of the subject insects.

Two types of geographical rarity are 'suffusive rarity' (species that are rare in special localities throughout their range) and 'diffusive rarity' (common at some locations, but rare at others). These are useful broad categories for conservation management. As insects are small and directly or indirectly tied to plants, botanical classification of rarity is of practical value. Various combinations of wide or narrow geographical range, (associated with broad or restricted biotope specificity) which in turn is related to 'somewhere large' or 'everywhere small' local population size, are conceptually useful. But fragmentation of the landscape and changes in insect behaviour and manifested biology may cause an insect species to shift category.

Conservation of rare insect genetic material is essential. Particular focus should be on those areas where endemism is high, particularly in the tropics, the southern hemisphere and on geologically long-established islands. It is also of general value to conserve typical species and sites as well as rare endemics. Additionally, conservation of species dynamo areas provides some protection for future biodiversity.

Although autecological studies are valuable in countries such as Britain, the loss of genetic diversity in the tropics through habitat loss and landscape modification is of the most urgent concern. If, conservatively, 1000 species are being lost per year, as many as 10 million alleles could be disappearing annually, although many of these would not necessarily be unique.

Possibly only 5–10% of all insects have been described, compared with over 80% of vertebrates. Although about 99% of the British insects have been scientifically described, the figure for tropical species is probably less than 1%. Even with this taxonomic impediment, Q. Wheeler's 'species-scape', in which different taxa of all biotic groups are illustrated in proportion to the number of described species, is an extremely potent message on the numerical importance of insects in ecosystem functioning numerical. This helps to overcome the mind-set impediment to recognizing the importance of insects.

The IUCN recognizes several categories of threat to species and a few selected biotic phenomena. In the USA, there are two categories. For insects in particular, assignment of species to a particular category is largely subjective. It may be more appropriate, but still difficult and subjective to use categories based on population viability. Again, for most insects their population behaviour and our lack of knowledge makes assignment difficult, especially with the impending global climatic changes overlying the local situations.

Despite the difficulties, category assignment in the *IUCN Red List of Threatened Animals* and in the *Red Data Books* (RDBs) focuses attention on rare and threatened taxa, albeit the larger and more conspicuous insects in general. The RDBs are often more politically delimiting than biologically realistic. Exemplary species that are listed may be threatened in one country, yet quite abundant elsewhere. Sometimes even localized introduced species are listed. Nevertheless, RDBs are valuable conservation management tools.

Atlassing reflects collecting sites as much as species distributional range. For the few countries that have fairly comprehensive mapping schemes of the most conspicuous groups, the maps are useful in illustrating range changes. In Britain, dots representing 10 km^2 give a good picture of the ranges of species of butterflies, dragonflies and orthopteroids. At finer scales, the insects occur in specific landscapes and biotopes within the marked area, as most insects are small with relatively small home ranges. This is especially so for the insect species of particular conservation concern. Elsewhere in the world, where the insect fauna is rich, mapping of specific species is logistically unrealistic, unless in designated reserve areas. Identification of plantscapes, biotopes and landscapes is an essential protection umbrella for the diversity of insects. Such identification maps involve the use of geographic information systems (GIS).

Part Three
Entomologists' Dilemmas

8

Insect pest control and insect conservation

The idea that insects merit conservation is not new, but it is one that the community at large finds difficult to accept. We have been conditioned to believe that insects should be slaughtered indiscriminately...

Tim R. New

8.1 BIOTIC CONTAMINATION BY ANIMALS

8.1.1 Invasive insects and other animals

Introduction of large mammals such as cattle and horses has led to landscape degradation. Myers (1979) coined the term 'Hamburger connection', to describe the rapid loss of Amazonian forests to ranching of cattle in order to gain economic returns by supplying cheap meat to the North American hamburger market. His notable account was an initial flag for conservation awareness of tropical forests. Not only has there been direct habitat loss, but the ranching has retarded ecological restoration.

Islands have been particularly severely affected by accidentally or deliberately introduced species of many taxa. Howarth and Ramsay (1991) comment on several lost insect species, including three large weevils that were probably eliminated in New Zealand by the Polynesian rat. While in Hawaii, three earwigs, *Anisolabis* spp., described from museum specimens, apparently are now extinct, perhaps as a result of the impacts of introduced animals of various types. Two Hawaiian moths, *Heliothis* spp., were annihilated by rabbit activity, and possibly also by sterile crosses with introduced relatives or by introduced parasitoids. In New Zealand, landscape modification and rat predation have seriously depleted populations of the Wellington speargrass weevil (*Lyperobius luttoni*), while mice are implicated in reducing populations of the range-restricted Stephens island weta (*Deinacrida rugosa*)(Figure 8.1). We will never know the full impact and extent of introduced predators. Glimpses of the picture are clear where small rat-free islands have retained their fauna, such as the New Zealand stilbocarpa weevil (*Hadramphus stilbocarpae*), while neighbouring rat-infested islands have lost it.

Invasive tramp species, especially certain ants, are notorious for their interspecific competitive abilities. The big-headed ant (*Pheidole megacephala*) is highly asymmetrically competitive (Samways, 1983a) and may be supplanting leaf litter insect species in Hawaii and New Zealand as well as species on other islands. The Argentine ant (*Iridomyrmex humilis*) is another aggressively harmful ant that has become highly invasive in South Africa (De Kock and Giliomee, 1989), affecting the limited and notably species-rich Cape fynbos ecosystem reserves (Giliomee, 1992). After displacing other ants (De Kock, 1990) that are beneficial by burying the seeds of indigenous myrmecochorous fynbos plants, they disperse such seeds themselves, but for only a few centimetres. They eat the elaiosomes without burying the seeds, which are then exposed to predators and the sun, with little chance of germinating. This impacts on the myrmecochorous fynbos plant community (Bond and Slingsby, 1984; Bond and Breytenbach, 1985), and will, in due course, adversely affect the dependent insects.

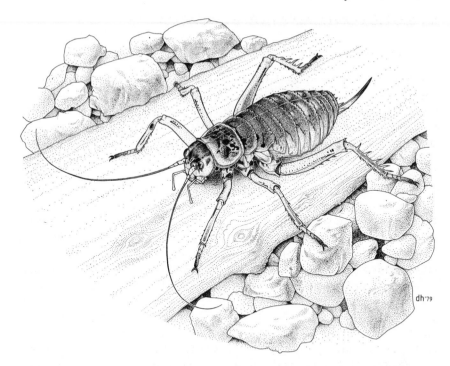

Figure 8.1. *Deinacrida rugosa* (Stephens Island weta). Confined to Stephens and Mana Islands, Cook Strait, and probably once widespread throughout mainland New Zealand. (Drawn by D.H. Helmore. From Howarth and Ramsay, 1991.)

8.1.2 Invasive insects and classical biological control

With increasing intensity of traffic along international trade routes, there has been an ever-growing invasion of alien biota in environmentally suitable new areas (Kornberg and Williamson, 1986; Macdonald, Kruger and Ferrar, 1986; Mooney and Drake, 1986).

In the USA alone, there are 1554 established insect invaders (Sailer, 1983), while 28% of all the insects on Tristan da Cunha are exotics (Holdgate, 1960).

As decades pass, crops around the world gradually accumulate the same or a similar complement of pests. Insect herbivores gradually establish themselves on the crop in one area after another, climate permitting. The result is that many worldwide crops are hosts to many worldwide pests. Many of these pests are therefore not native to the land in question, and, in the absence of natural biotic-suppressive agents, increase to high and economically damaging levels in their new homes. Although only about 200 or so insects are really serious pests, their impact is enormous, with probably at least 30% of all crop harvests worldwide lost to them. The actual number of pests is much higher than this, with some estimates at 10 000 damaging insects and mites. Even so, this figure is only about 1% of described species.

These invader pests are often targets for classical biological control (CBC). The principle involves the reduction of an exotic pest population by selected exotic natural enemies imported for this express purpose. It attempts to employ specific agents against specific target pests, ideally without impinging upon any external abiotic or biotic factors. The desirable outcome is that the pest is reduced to, and maintained at, a level well below the economic threshold (e.g. DeBach, 1974; Samways, 1981) (Figure 8.2).

When a pest is exotic, it is an accidentally introduced, highly significant (biologically and economically), biotic pollutant or contaminant. Such contamination is virtually irreversible. Introduction of a biocontrol agent is further, irreversible contamination, although unusual in that it reduces the level of the first pest contaminant when the programme is successful (Samways, 1988b) (Figure 8.3). When the programme is unsuccessful, it is because the introduced natural enemy dies out, or establishes but is not economically effective. In conservation terms, the first outcome results in no contamination. The second outcome is the one of greatest concern, because

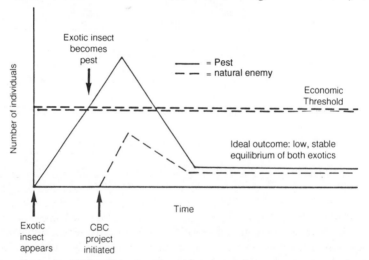

Figure 8.2 Illustration of the principle of classical biological control (CBC) when an insect accidentally appears in a new area and, without its native natural enemies, increases to high population levels where it becomes a pest on certain plants – usually ones of horticultural or agricultural significance. A **selected** natural enemy is then collected in the native home of the pest, quarantined, reared and released against the pest. The intended outcome is that the natural enemy disperses and increases to such an extent that it reduces **only** the target pest. If all goes well, the two then remain at a low, stable equilibrium, with no impact upon any other native fauna. (From Samways, 1988b.)

of the inherent risk that the biotic contamination may be competitive with, or attack, non-target native organisms.

In the past, there have been forceful debates on the harmful impact of pesticides on the environment (Carson, 1963; Mellanby, 1967). In contrast, CBC has generally been considered, until recently, the ideal 'natural alternative'. Three decades ago, the debate was often polarized, with opponents rarely seeing the balanced advantages and disadvantages of both alternative control methods. In reality, many crops involve integration of pesticides and biocontrol, with pest management generally concentrating on reducing pesticides that may be damaging to natural enemies. Until recently, little consideration was given to any environmental impact that either control method may have had on rare insects outside the crop environment.

8.2 PESTICIDES

Pesticides in the past have been relatively cheap, easy to apply, readily available and the impact on the pest visually impressive. Pesticides are now no longer cheap, new types of compounds are becoming extremely scarce,

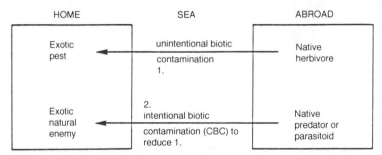

Figure 8.3 Illustration of the concept of biotic contamination. A herbivore, which need not be a pest in its native land, becomes accidentally established in a new 'home area'. In this new land, its population levels may soar in the absence of its natural predators, so that it becomes an exotic pest in the new home area, and there constitutes an unintentional biotic contamination. To reduce this new pest in its new home area, a search is made overseas for an effectively suppressive predator or parasitoid natural enemy. The natural enemy is then deliberately, carefully, and selectively, introduced into the new land (classical biological control, CBC). Although this exotic natural enemy constitutes further biotic contamination, it has the great advantage of specifically reducing the level of the pest (namely the 'unintentional biotic contamination 1'), thereby reducing the overall biomass of biotic contamination in the new area. An important point is that both the unintentional and intentional biotic contaminations are virtually irreversible. This implies thorough screening before release of the biocontrol agent to lessen any risks that the natural enemy will not extend its host range to any native insects once released. (From Samways, 1988b.)

and the environmental impacts are well known. Some compounds, especially some of the organochlorines are bioaccumulative (Moore, 1987a), having major impacts on some predators of insects, especially raptors, well along the food chain. Others, particularly soil-applied systemic carbamates, find their way into water courses. Despite these well-known risks, some persistent pesticides are still widely used in tropical areas. Organochlorines are used extensively for mosquito control, and in the neotropics for leaf-cutter ant suppression.

Having made these points, unquestionably, pesticides can fragment, reduce and even eliminate local populations of target and non-target organisms. But there is apparently no conclusive evidence that a pesticide has definitely been primarily responsible for an insect species extinction. However, there is a note of caution, in that disappearances are difficult to detect, and pesticides in waterways may have been a contributing rather than a primary cause of extinction. An important point is that insect populations that have been reduced can recover, even though it may take several years after the pesticide perturbation. Pesticide application can be terminated immediately, with relatively little impact momentum. This contrasts with the higher impact momenta of CBC, landscape modification, habitat loss and global climate change, the effects of which are difficult to halt let alone reverse. Nevertheless, pesticides should still be used with caution, as they can cause relative changes in species abundances and community structures, so influencing ecosystem processes.

Pesticide resistance must also be taken into account, and 500 or so insect and mite species now are resistant to pesticides. Resistance is a relatively permanent phenomenon, with the genes remaining in the population at least for several dozens of generations. These resistant pests have a competitive advantage in ecosystems over other species that have had their populations reduced through pesticide pressure.

8.2.1 Some comparisons of biological and chemical control

CBC, unlike pesticides, cannot be halted once an agent has been released. Although intentionally target specific, it has the problem that once an agent released it cannot be recalled. Inundative release is the continual release of large numbers of natural enemies, often against indigenous pests. In some respects, its application is more like pesticide usage in that its impact is on a repetitive basis that generally wanes following repeated 'applications', leaving a 'residue'. The disadvantage of inundative release is that the natural enemies, when not highly host specific, as in the case of egg parasitoids of the genus *Trichogramma*, can attack on-site or on nearby non-target organisms.

Overall, the entomologists' dilemma is that pesticide usage, CBC, inundative release of natural enemies and certain cultural practices all have some harmful effect on insect conservation. These biotic impacts are additive upon the effect of the fragmentation of the landscape by agricultural enterprise. In addition, all methods, particularly dispersive and highly multiplicative biocontrol agents, extend their impact and contamination well beyond the agricultural patch boundary. The economic benefits must be weighed against environmental health. This requires sustainable agriculture and the use of agroecological techniques that are more in tune with temporal and spatial scale processes of natural systems. Putting the ideas into practice is now the urgent and cooperative task of agriculturalists and conservationists.

8.3 CLASSICAL BIOLOGICAL CONTROL – REALISTIC ADVANTAGES

The advantages of general biological control by encouraging enhanced natural predation and parasitism through landscape modification for sustainable agriculture was discussed in sections 6.4 and 6.5. Here, the emphasis is specifically on CBC. General texts on CBC have cast the approach in a highly favourable light, with some failures but also some outstanding successes (Caltagirone, 1981; Beirne, 1985) at the sole expense of the pest (Caltagirone and Huffaker, 1980). The methodology has appeared environmentally clean in comparison with pesticide control methods (Mellanby, 1967).

Until recently, CBC has generally been well received by environmentalists and conservationists. CBC, when properly carried out by responsible agencies, uses host-specific natural enemies in agricultural landscapes, while conservation management has aimed to protect threatened species in particular biotopes. These protected species have generally been highly localized, with low reproductive potential and very particular biotope requirements. Many pests, in contrast, are relatively eurytopic and of high reproductive capability. Even the listed threatened species (e.g. Wells, Pyle and Collins 1983) are generally of quite different taxa to those of pests, and include threatened dragonflies, damselflies, stoneflies and earwigs. But this is not to say that biocontrol target taxa do not contain threatened species.

The notable CBC successes have involved arthropod hosts of a particular range of taxa with little overlap with threatened species. The outstanding successes listed by Caltagirone (1981) are small and numerous, including a spider mite, a whitefly, a stink-bug, an aleyrodid blackfly, two aphids, three scales, a mealybug, a moth and a beetle. Only the last, the rhinoceros beetle (*Oryctes rhinoceros*), is large and conspicuous, albeit with a relatively large reproductive potential.

These successful biocontrol programmes have brought enormous economic returns, and from a human viewpoint this has been extremely important, especially under poor socioeconomic conditions where pesticides may not be feasible, affordable or available.

There is some evidence that CBC can be favourable to conservation where a target herbivore is causing damage in a nature reserve. The natural enemy may well disperse and locate a pest that has spread to conserved areas. Many exotic scales and mealybugs infest the rich endemic plant fauna of South African indigenous plants. The cosmopolitan red scale, *Aonidiella aurantii*, infests trees such as *Trichilia* spp. and *Rhus* spp., but is generally kept at low levels by a combination of the introduced parasitoid *Aphytis melinus* and the partially introduced coccinellid *Chilocorus nigritus,* as well as by the indigenous *Aphytis africanus*, this last species being an interesting case of a highly influential indigenous agent suppressing an exotic pest (Samways, 1988a).

Apparently, the only major biocontrol programme undertaken for the express purpose of conservation was the attempt to control the diaspid scales, *Carulaspis minima* and *Lepidosaphes newsteadi*, which invaded the island of Bermuda in the early 1940s and devastated the natural forests of Bermuda cedar (*Juniperus bermudiana*). These forests were part of the natural heritage of the island, and to save them, several natural enemies, including the ladybird *Rhyzobius lophanthae* and an unknown parasitoid, were tried. The natural enemies did not control the pests quickly enough. In 1951, the project was terminated, when most of the trees were dead or dying. Initial misidentification of *C. minima,* the impact of hurricanes and predation of the ladybirds by lizards were all thought to contribute to failure of this expensive project. The Bermuda cedar has not become extinct, although it is not known whether the natural enemies did in fact avert its total demise (Bedford, 1949).

In biocontrol circles the term 'conservation' is used to describe the maintenance of both native and introduced natural enemies (DeBach, 1974). Unnecessary or potentially harmful pesticides or pesticide-application methods are avoided. Herbaceous strips of vegetation with nectiferous plants as feeding and nesting places for natural enemies are established. These approaches are generally beneficial to insects and other animals, beyond the natural enemies themselves.

Most importantly, CBC and other methods of insect control are intended to benefit man. And CBC, at least for certain pests, has indeed been highly advantageous to man's well-being, which is central to *Caring for the Earth* (IUCN, 1991). It is a question of relative impact, and the record of benefits to man that CBC has accrued cannot be denied. It is almost certain that modern CBC activities with an emphasis on monophagous rather than polyphagous agents, and with strict quarantine measures, is probably far less detrimental to insect diversity than landscape degradation and the accidental introduction of invasive insects. But the history of biological control has by no means been without harm to indigenous faunas.

8.4 BIOLOGICAL CONTROL – DISADVANTAGES

8.4.1 Vertebrate and non-insect agents

Many long-past projects, especially those involving vertebrate and some non-insect invertebrate agents, have had disastrous consequences upon some indigenous faunas. The record for snail agents has been exceptionally poor in the conservation context (Anonymous, 1983). The fish *Gambusia affinis,* although effective against mosquito larvae, has been found to be damaging to populations of zooplankton and other fish species, as well as possibly causing massive population reduction if not extinction of the damselfly *Megalagrion pacificum* in Hawaii (Moore and Gagné, 1982). The toad *Bufo marinus*, which was introduced into Australia to control scarabaeid larvae, *Dermolepida albohirtum*, in sugar-cane, has had little impact, and the agent itself has become a nuisance and a predator of many non-target insects (Wilson, 1960). Today, the use of vertebrates, molluscs and any other highly competitive, polyphagous potential biocontrol agent is rarely even considered by bona fide biocontrol agencies.

8.4.2 Classical biological control and extinctions

Occasionally, CBC projects do cause extinctions, especially within the geographical confines of small islands. The exotic citrus psylla, *Trioza erytreae*, was completely eliminated on Réunion Island by the introduced parasitoid *Tetrastichus dryi*. Of conservation note, this parasitoid switched to survive on an alternative, native host, *Trioza litseae (eastopi)*(Aubert and Quilici, 1983).

8.4.3 Pathogenic agents

Insect pathogens in the broadest sense include entomophilic fungi, protozoans, bacteria, viruses, rickettsiae and nematodes. Certain fungi (e.g. *Metarrhizium anisopliae*), protozoa (e.g. *Nosema locusteae*), bacteria (e.g. *Bacillus thuringiensis*), viruses (e.g. of some lepidopterans) and nematodes (e.g. *Steinernema feltiae* = *Neoplectana carpocapsae*) have been employed in some localities against specific target pests, although many have an actual host range greater than that of the target species.

These are pathogens that occur naturally. For use in insect suppression they are cultured in the laboratory and applied in high concentrations against the target pests on a specific crop, or area of forest or rangeland.

These insect pathogens are not to be confused with plant pathogens, which may be carried by herbivorous insect vectors. Biocontrol agents may inadvertently carry plant pathogens acquired in the vicinity of their herbivore hosts. During a CBC programme, these are screened out, along with hyperparasitoids, in quarantine. Some animal pathogens may also be screened out. The foot-and-mouth virus of cattle is removed through formalin sterilization from scarabaeid beetle eggs dispatched from South Africa to Australia in attempts to use the dung beetles to bury Australia's vast number of cattle dung pats.

Insect pathogens have become increasingly widely used as chemical pesticides in insect pest management programmes. But pathogens are variably infective to different insects, and their more widespread use must be monitored closely for their impact on insect diversity.

Bacillus thuringiensis israelensis, which is used in mosquito control, has definitely caused mortality of mayfly and dragonfly larvae and other aquatic insects (Zgomba, Petrovic and Srdic, 1986). To date we have only glimpses of evidence of the impact of pathogens on non-target organisms, and it is undoubtedly a field for much further research. Whether or not some of the pathogens that exist in any one insect in one area are native may be difficult to determine. This is especially the case after so many deliberate and accidental insect introductions, which have unwittingly involved the carrying of pathogens.

Insect pathogens are biopesticides. They are living organisms that can be manipulated genetically to improve their efficacy. One group of pathogens that is receiving such attention is the nuclear polyhedrosis virus (NPV) subgroup of baculoviruses, which has been considered to be sufficiently specific to be registered, for example, for control of the bollworms *Heliothis* spp. Genetic modification of NPVs by cloning toxin-encoding genes into NPVs has increased their virulence (Hammock *et al.,* 1990; Tomalski and Miller, 1991; Stewart *et al.,* 1991). But doubts have been raised about the new host-range specificity. The assumed host specificity of NPVs expands when there is co-infection of two unrelated NPVs in moths such as the alfalfa looper (*Autographica californica*) and the silkworm moth *(Bombyx mori).* New viruses can form by extensive crossover of large regions of DNA, and these new forms are then able to attack a wider range of hosts (Williamson, 1991). Of particular concern is that, if recombinant viruses are applied at high concentrations, this may result in the formation of new, unpredicted NPVs (Hochberg and Waage, 1991). The biopesticides in recombined form may well then become pests themselves (Altmann, 1992) both against non-target native fauna and against already effective and safe natural enemies.

8.4.4 Contentious issues

Howarth (1983, 1991) and Samways (1988b) have warned of the wider environmental implications of CBC on indigenous insects. The historical context is important here. Until the recent awareness of nature conservation and the magnitude of the biodiversity crisis, CBC projects were aimed at specific targets without any real concern for environmental consequences. The first CBC programmes were in existence long before the advent of organic insecticides and, during the polarized debates of the 1950s and 1960s between chemical and biological control practitioners, insect conservation was all but forgotten. The only real concern was conservation for the effective biocontrol agents in the face of pesticide applications in the economic context.

Today the leading CBC agencies are well aware of possible repercussions of their introductions on indigenous faunas. Caution is exercised, with highly polyphagous predators banned from introduction into some countries. But Howarth's (1991) warning of the risks of the casual trade in 'beneficial insects' and brandished euphemistically as 'using nature to fight nature' is highly relevant, and of very great insect conservation concern.

Actual evidence of a properly planned CBC project irretrievably damaging native faunas is difficult to ascertain. Any real **proof** is almost non-existent. This is not to say by any means that there has not been damage to native insect faunas by introduced agents. It may in fact be more widespread than commonly thought, but cryptic in effect.

It appears that the earlier more indiscriminate projects were the most damaging. Zimmerman (1958) mentions, but does not name, introduced parasitoids that have damaged native Hawaiian lepidopteran fauna. Also, Taylor (1979) has suggested that the decline of some indigenous Australian aphid species is due to the release of parasitoids for control of exotic aphids.

Howarth (1991) lists other possible extinctions through exotic introductions. Several Pacific island Lepidoptera and Hemiptera may well have become extinct as a result of introduction of both tachinid and hymenopteran parasitoids. The historical context must be re-emphasized, and the *laisser-faire* approach by earlier workers towards indigenous faunas is totally unacceptable in today's context.

Some natural enemies have been instrumental in reducing and then causing extinction of either target or non-target hosts. The extinction itself may have come about through synergistic action between the biocontrol agent, host habitat destruction and landscape modification, rather than simply the effect of the agent. Nevertheless, the influence of an agent can be substantial. The levuana moth (*Levuana iridescens*) became extinct on Fiji following the introduction of its parasitoid the tachinid *Bessa remota*, from Malaysia in 1925. The agent eliminated the target by the 1940s, and today the tachinid continues to parasitize other non-target hosts (Howarth, 1991).

CBC is still largely empirical, and even if the agent does not establish on the host it may remain on alternative hosts. Such environmental impacts are not usually considered by biocontrol practitioners, especially as it is a 'failed' project, with little or no economic future. Such instances are probably rare, because projects generally track the progress of the introduced agent. Pretesting before release is rare for entomophagous agents, but is regular practice

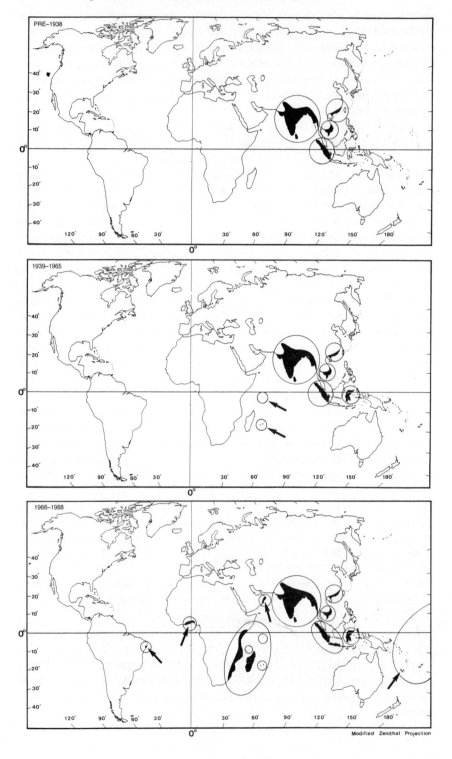

among weed biocontrol specialists, principally for the economic reason that the herbivore may turn to feed on a crop plant.

Caution is definitely warranted. The polyphagous coccinellid and biocontrol agent *Coccinella septempunctata*, although commonly an aphid feeder, will feed, both as a larva and adult, on the eggs and first-instar larvae of the threatened North American lycaenid butterfly *Everes comyntas*, when aphids are unavailable (Horn, 1991).

8.4.5 Risks with polyphagy and monophagy

The biggest risk with a biocontrol agent is that if it is polyphagous it may readily reduce a non-target organism to precariously low population levels. This low density may make mate-finding difficult, make local extinctions more likely when weather conditions are harsh, and may even result in detrimental narrowing of the genetic variation. As host switching by introduced agents poses risks to indigenous faunas, polyphagous natural enemies are avoided as much as possible. Indeed, it is polyphagous, eurytopic agents that have given the most problems in the past.

Climate and host range are important in determining the ultimate established range of a natural enemy (Franz, 1964; Samways, 1989d) (Figure 8.4). This range is often less than that of its host. The limitation of climate is the same for a monophagous and polyphagous herbivore, although inevitably there will be more widespread, suffusive establishment of a natural enemy on a polyphagous than monophagous herbivore, as the collective host plants will be denser and have a greater range.

The links and lengths of food chains may be significant, especially in relation to diet width (Pimm, 1987). This may be important for conservation when a natural enemy that removes a predator of a herbivore is introduced. For a monophagous herbivore, a possible outcome is that the **plant** popula-

Figure 8.4. The increase in biogeographical range of a highly invasive natural predator and biocontrol agent of scale insects, the ladybird *Chilocorus nigritus*. A native of the Indian subcontinent and Indo-China, it was deliberately introduced in the late 1930s into Mauritius and the Seychelles (centre map, arrows). Since then it has spread to neighbouring islands and mainland Africa. There has also been an eastwards expansion into several Pacific islands. Recently the beetle has also appeared in north-east Brazil, West Africa and Oman (lower map, arrows). There is good circumstantial evidence that this spread has been mostly due to hitchhiking along shipping routes. Although the species is a good invader, it has specific climatic tolerances. Inland it is restricted principally to H. Walter's zonobiome II and in coastal areas to zonobiomes I–II. Its appearance in Oman is anomalous, and perhaps indicates a highly local suitable climate, or an indication that the species' climatic tolerances have increased. The dilemma is that this is a highly valuable predator/biocontrol agent of some serious scale pests, yet it is highly invasive, and in southern Africa, may have displaced *C. wahlbergi*. *C. nigritus* can be expected in future years to spread further throughout the tropics and subtropics. (From Samways, 1989d.)

tion is reduced, and if the plant itself is endangered the monophagous herbivore could be reduced to extinction. But there appears to be no such examples to date. Removal of a predator of a polyphagous herbivore could possibly see the demise of the most preferred plant hosts, while nevertheless allowing survival of the most resistant or least-preferred host.

8.5 RESOLVING CONFLICTS

In reality many of the classical biological control (CBC) projects have principally benefited large and small farms, estates or plantations rather than the subsistent farmer, who may rely more on crop rotation, mixed cropping, small fields, natural predation and other agroecological approaches. This is not to generalize, because all types of farming communities can benefit to a greater or lesser extent from both natural predation and the outcome of successful CBC projects.

Strict quarantine measures are an essential part of a CBC exercise. Of the invaders, most have arrived of their own accord, with or without the deliberate agency of man. Even stricter quarantine is needed now that insect genetic resources are being depleted at such a rate. What will eventually be left will be a depleted genetic pool. In turn, there is the possibility that genetically engineered CBC agents may pose even greater threat, unless the modifications are very specifically aimed at economically important targets that are subsequently maintained at a low, stable equilibrium, with no impact upon non-target organisms.

CBC clearly has an added responsibility to biodiversity conservation that was not realized in earlier years, particularly by practitioners whose attack was focused on pesticide usage. It is essential now that CBC becomes more predictive, with greater knowledge of the exact short- and long-term effects, to avoid repercussions that reduce insect diversity. To this end, screening for specificity and for the carrying of potentially harmful pathogens is becoming increasingly important. Also, an added responsibility will be associated with genetically engineered biocontrol agents. There are also legal implications because although natural organisms are not patentable, modified organisms can be patented. This automatically implies liability. Although these cautionary activities will shackle CBC to some extent, it also emphasizes the need to use sustainable agroecological techniques that are genuinely realistic to both the farmer (who sustains us all) and to preservation of insect diversity.

8.6 INVASIVE PLANTS AND BIOLOGICAL CONTROL OF WEEDS

8.6.1 Weeds and insect conservation conflicts

'Weed' is a human concept, being a plant growing where it is not wanted.

Figure 8.5 A weed is a plant growing where it is not wanted. *Lantana camara* is a highly noxious and invasive plant in some areas and a major target weed for biological and non-biological control. In other areas, it is planted in 'insect gardens' as it is highly attractive to many nectar-feeding insects, especially butterflies. (From Cilliers and Neser, 1991.)

Many weeds are economically insignificant. These are the early successional species that are anathema to horticultural landscapists. These are the pioneer species of bare ground, which carry their characteristic complement of early successional insect herbivores. Many of these pioneer plants are prolific flowerers and are attractive to nectar-feeding insects, especially eurytopic butterflies.

There are some contradictions in what is a weed and what is not. Plants such as *Buddleia davidii*, *Cosmos bipannatus* and *Lantana camara* are recommended as attractant nectar sources in ecologically landscaped butterfly gardens (Booth and Allen, 1990). Yet in some geographical areas and situations, where butterfly gardening is not a priority, these plants are weeds. *L. camara* can be particularly serious, as it can be highly invasive, competitive of native plants and toxic to cattle, forming impenetrable thickets (Figure 8.5). In some countries it is a declared noxious weed subject to legislation and a major target for an intensive biocontrol programme (Cilliers and Neser, 1991). Such conflicts need to be resolved in the spirit of a future sustainable world. But there are various degrees of balance. The Chinese *B. davidii* was introduced into Britain in the last century and has been planted to attract butterflies such as the peacock (*Inachis io*), comma (*Polygonia c-album*), brimstone (*Gonepteryx rhamni*) and small tortoiseshell (*Aglais urticae*), all relatively common butterflies. The plant has since spread of its own accord and is common today in many disturbed sites such as railway sidings. Here *B. davidii* produces a pleasing inflorescence and creates butterfly activity. It is only a weed when it is invasive to an already aesthetic area.

8.6.2 Invasive weeds

Worldwide, hundreds of plants have been accidentally or deliberately translocated and established in new, climatically suitable lands. Some of these, both terrestrial and aquatic species, have become highly invasive, altering community structure and damaging native ecosystems. Some weeds, such as *Chromolaena odorata*, a native of Brazil, can grow so rapidly as a creeper that they smother native vegetation in the subtropical forests of southern Africa. Similarly, some of the Australian eucalypts, although grown commercially in southern Africa, can also establish outside forested areas. The intensive leaf-full and dense canopies create a carpet, which is highly susceptible to combustion. This frequently happens naturally, as well as through human action: lightning strike frequency in Natal is among the highest in the world at 10 strikes km yr^{-2}.

Weeds, by altering the plant community pattern, also alter the insect community, both on the plants themselves and in the soil below. Some polyphagous native insects, may, in turn, extend their host range by beginning to feed on the new host plant. This extended behaviour is regulated by plant architecture as well as by plant chemistry (Moran, 1980). But generally these invasive plants reduce local biotic richness.

Often the spread reaches beyond the boundaries of high economic value agricultural and urban land, and extends into natural and seminatural landscapes. Nature reserves and wilderness areas are also susceptible. Water weeds spread rapidly along water courses, or are transferred from one standing water body to another by bird movement. Dams, lakes and rivers may become clogged. It is not always easy to distinguish between direct and indirect effects. *Pinus roxburghii* depresses grasshopper richness under its canopy (Samways and Moore, 1991). Several factors seem to interact. The pine-needle mat inhibits native plant growth, and also acidifies the soil (Boyd and Marucci, 1979), while the canopy reduces light levels for understorey plant growth and for grasshopper development.

In some cases, the weeds create physical conditions that alter the local distribution pattern of insects. The influence may extend beyond the drip zone of the plants. The influence may not always be suppressive. *Cupressus arizonica* stands enhance some grasshopper populations in the neighbouring grassland (Samways and Moore, 1991).

Chemical or mechanical control of invasive, widespread weeds is not often logistically or economically feasible, as the weed density may be too high, the terrain too remote or the general cost of control too high. A biocontrol programme, using carefully selected, quarantined and screened herbivorous agents, may be the only viable option. These natural enemies disperse, seek and suppress or destroy local populations of the target host plant.

Many of these exotic weeds are rare plants in their native countries, suggesting that they are normally suppressed by herbivores, seed predators or pathogens or through competitive interactions with other plants. Where herbivores are key suppressers, these plants, which are then weeds in the new country, are prime subjects for biocontrol by introducing the absent herbivores.

Over 87 species of terrestrial weeds (Julien, 1987) and several species of aquatic weeds have been targets in biocontrol programmes. These have involved about 100 specific insect herbivores, with about a 30% success rate (Waage and Greathead, 1988). These projects have been primarily for economic reasons, for pastures, forestry and crops. Some of these programmes have been enormously successful, as with control of the prickly pear cactus *Opuntia stricta* in Queensland after introduction of the moth *Cactoblastis cactorum* from Argentina. Many projects have been only partially successful either in area or in level of suppression. Biocontrol of crofton weed, *Ageratina adenophora*, with the tephritid fly *Procecidochares utilis* and the fungus *Phaeoramularia* sp. in South Africa has been only partially successful because the moist biotopes of the weed do not favour the fly. Interestingly, the fly is also host to native parasitoids (Kluge, 1991), so increasing their density.

Some of the programmes against water weeds have also been highly successful. For example, the floating plant *Salvinia molesta*, a native of Brazil, has invaded and since been controlled in several Far Eastern and African countries, principally by the weevil *Cyrtobagous salviniae*. But even invasive water weeds can be beneficial to native insect fauna. *Pistia stratiotes* has invaded and clogs part of the Sabie River system in the Kruger National Park, South Africa, yet hosts high populations of some native Odonata (Clark, 1991).

Successful weed biocontrol programmes are, in general terms, beneficial where the philosophy is management of the landscape to a near-pristine condition, or at least to a biotic structure as near as possible to its condition prior to human influence.

Entomologists' dilemmas are particularly acute in the case of weed biocontrol, especially as one man's weed is another man's income. Insects are being translocated which cannot be recalled, and have variable potentials for feeding on other plants whether crop or non-crop. The possible impact on crops that are not targets is obvious, and screening programmes minimize the possibility of wrong targets being attacked. Also of concern, and not usually tested, is the risk that a *Red Listed* threatened plant might be attacked. The risks are probably low, considering the low apparency of threatened plants and that the taxonomic remoteness from the target weed is often great (with concurrent effective defence systems against non-adapted herbivores). But this is not to encourage complacency. Loss of a threatened plant is not only a biotic loss *per se*, but also a loss of its associated insect community.

8.6.3 Conflicts with weed biocontrol

Entomologists are at the centre of weed biocontrol conflicts, working to benefit mankind and nature conservation, yet at the same time running the risk of making an error in insect (or mite, or pathogen) translocation.

The use of genetically engineered organisms for weed biocontrol or for insect biocontrol is imminent. But it is essential that there is thorough checking for possible rogue agents before release, particularly if genetic stability cannot be guaranteed. Of direct concern is that a large genome is being manipulated. The potential ramifications of altering genomes comprising several thousands of genes are enormous, although, interestingly, it seems that in nature genomes are not necessarily always stable within one species, being modified by outside agents. Houck *et al.* (1991) have found strong inferential evidence for the transfer of genes by a semiparasitic mite *(Proctolaelaps regalis)* from a species of the *Drosophila willistoni* group to *D. melanogaster*. Transfer of genetically modified DNA segments could disastrously extend the host range of a herbivore, particularly if driven by a strong selective advantage.

Another cautionary note is that insect pathogens are widespread and often highly contagious. Microsporidian protozoan pathogens are common in insects, often causing chronic diseases reducing overall fitness, particularly when the insects are under stress. Microsporidia do not generally prevent establishment of the biocontrol agent, but they can suppress populations of their hosts. Although most microsporidians have a narrow host range, there is always the risk that release of herbivorous biological control agents will transmit disease to indigenous insect fauna (Kluge and Caldwell, 1992). There is never any guarantee that no harm will be done to the indigenous fauna. This is of major proportions considering that pathogens such as *Nosema* spp. are notoriously difficult to cleanse from the biocontrol cultures.

There are other aspects of weed biocontrol that require caution. The prickly pear *Opuntia ficus-indica* is a native of Mexico and once infested 900 000 ha in arid parts of South Africa. The cochineal insect *Dactylopius opuntiae,* along with the moth *Cactoblastis cactorum,* was introduced in the 1930s and has reduced the infestation area to about 100 000 ha. In the intervening years there have been second thoughts regarding the biocontrol agents, which are now regarded as pests. This is because *O. ficus-indica* is now used as drought fodder for stock, as a fruit and vegetable source for humans and as a host plant for the commercial production of the red dye, carminic acid, for certain foods, cosmetics and pharmaceutical products (Zimmerman and Moran, 1991). From an insect conservation viewpoint, the introduction of *D. opuntiae* not only partially restored the landscape but was beneficial to certain native predator populations, especially the coccinellid *Exochomus flaviventris.* Another twist in the story is that the ladybird *Cryptolaemus montrouzeri,* which was introduced to control the citrus mealybug, *Planococcus citri,* has since used *D. opuntiae* as an alternative host.

From these examples, the translocations of insect populations for weed biocontrol clearly have beneficial aspects for man and native insect communities, but there are also potential consequences that are extremely difficult to predict prior to release of the agents.

8.7 SUMMARY

Deliberate and accidental introduction of mammals to new areas has caused landscape degradation and predation of insects, and the demise of some localized insect species. Even some invasive, competitive insects, such as certain ants, have themselves caused other insects to be displaced. Worldwide, insect invaders whether harmful or neutral in impact, continue to accumulate, with over 1500 established insect invaders in the USA alone.

Although pesticides fragment and suppress local populations of insects and other animals, there appears to be no conclusive case of a pesticide being the prime causal reason for an insect species extinction. In contrast, invasive and some deliberately introduced organisms, especially vertebrates, are known to have caused insect species extinctions. Such impacts have occurred alongside biotope and landscape disturbance and degradation. This is not to condone pesticides which are well known to have had variable environmental impacts.

Exotic pests are a form of accidental biotic contamination. Classical biological control (CBC) is also a form of biotic contamination but unusual in that, when successful, it reduces the level of the pest contaminant. CBC has had some outstanding successes against some major pests on crops. Additionally, there are cases where invasive insects in nature reserves have been controlled. When the biocontrol agent is carefully selected to be highly host specific, there is considerable merit in the use of CBC.

The history of CBC has included some calamities, especially where non-arthropod and vertebrate agents have been introduced. Although insect parasitoids and predators have scored the most successes, there is evidence that even some of these have been harmful to native faunas, particularly in the Pacific region. Even a species of moth once common on Fiji appears to have become extinct through deliberate introduction of a tachinid fly agent, although habitat loss may also have been the major factor in its demise.

The use of mass-cultured, naturally occurring pathogens has been incorporated into some pest management programmes. These pathogens have been considered in some circles as environmentally safe alternatives to pesticides. However, the target host range is wide for many of these pathogens. Further, the new generation of genetically engineered nuclear polyhedrosis viruses poses some serious threats, as the host specificity cannot be guaranteed.

'Weed' is a human concept, being a plant growing where it is not wanted. Some horticultural annual weeds are simply early successional plants that support their characteristic insect community, as well as providing nectar for tourist insects. Some weeds are even planted to attract butterflies yet are declared noxious weeds elsewhere. Other planted exotics, especially pines and eucalypts, can be particularly suppressive to local insect populations.

Invasive weeds can be highly damaging to some native plant communities, through smothering and increasing the risks of combustion. Such weeds, as with plantations, reduce local insect species richness, sometimes beyond the drip zone of the trees.

Biocontrol of invasive weeds has been about 30% successful, although some weeds have been distinctly beneficial for some native aquatic and terrestrial insect species. There are always risks with introducing weed biocontrol agents, that they may expand their host range to other plants, both crop and non-crop. Further, insect agents frequently also carry insect pathogens which are difficult to rid from cultures and may readily infest native insects.

There are some clear principles arising from past and present insect pest control. Introduction of generalist feeders is to be avoided. Strict quarantine is essential, with careful screening out of potential hazard organisms. Use of genetically engineered agents, especially pathogens, must be viewed with great caution. Unquestionably, if pest management can be achieved by landscape modification to promote insect diversity, then this is the best approach. Environmentally safe pesticides and carefully selected CBC agents can be used alongside cultural and landscape control methods. Besides maintaining crop yield on a consistent basis, the aim of pest control in the conservation context is to avoid reduction of the genetic diversity of the insect fauna, whether widespread and common species, or localized, taxonomically remote endemics.

—9—

Insect conservation ethics

For man has closed himself up, 'till he sees things thro'
the chinks of his cavern.

William Blake

Our role is to live out a spacetime, placetime ethic,
interpreting our landscapes and choosing our loves
within those landscapes. We endorse the world with
our signatures. In this sense we want an emotive ethic
but not, as that term usually conveys, an ethic that is
nothing but emotion. Emotive environmental ethics
lives in caring response to the surrounding natural
places and times, an appropriate fit of the tripartite
mind – reason, emotion, will – creatively responding to
the nature in which mind is incarnate.

Holmes Rolston, III (1988)

"No doubt, Sir, an Entomologist ?"

(from *Acheta domestica* (1849))

9.1 VALUE OF THE INDIVIDUAL INSECT AND THE SPECIES

9.1.1 Intrinsic value of the individual

Humans, in general, place little, if any, value on the life of an individual insect (Lockwood, 1987), and even on insects in general. Yet insects feature in extremes in ethical arguments. Colloquial definitions include a sadist 'who pulls wings off flies' and a humanitarian as 'one who wouldn't hurt a fly'. Sentience, the consciousness of an organism, includes pain, thought and awareness. Dawkins (1989) draws on William Blake's poetic identification with an insect:

> Am not I
> A fly like thee
> Or art not thou
> A man like me?

(Songs of Experience)

We all kill insects, some of us thousands each year, as vehicle splatter from night driving. If 17 million humans have been killed by the motor car since Karl Benz first introduced his prototype over a century ago, the number of insects that have died by the same agent is quite incalculable. Ironically for the insect, evolutionary years of predator avoidance behaviour has permitted it nocturnal activity, only to be eliminated by our own new nocturnal hyperactivity.

Indirectly you and me are also exterminators of insects in a different way. We like to see unblemished food in the supermarkets, pressurizing the farmer to be rid of all that may aesthetically detract from or destroy his produce.

What do we do about preserving the life of an individual insect ? Who would, as Albert Schweitzer did, prefer to keep the window shut and breathe a stuffy atmosphere rather than see one insect after another fall with singed wings upon his table?

Lockwood (1987) proposes a minimum ethic:

> We ought to refrain from actions which may be unreasonably expected
> to kill or cause non-trivial pain in insects when avoiding these actions
> has no, or only trivial, costs to our own welfare.

Mosquito nets on windows would prevent wing burning and at the same time be beneficial to human welfare. We can release trapped insects, drive our vehicles more slowly and spray less insecticide in our homes. The practical reality is that our general disrespect for insects (unlike that for many large vertebrates) clashes with Lockwood's ethic. Subjectivity also plays a role. Some people would, and others would not, decry the vivisection of a large unanaesthetized bush cricket for scientific neurophysiological experimentation. Similarly, we would readily open a window to release a fluttering butterfly, but we would not so easily do the same for a fly. Yet the two have almost identical nervous and sensory systems. Poor fly should there be any hint of it being even minimally harmful to our welfare, or even to our housepride, and especially if it is small in size. When an insect is inherently an imposition, it is spurned, and often physically squashed, or downed with an insecticidal spray.

Preservation of insect diversity lies in the intrinsic value of the individual organism in addition to the anthropocentric value of diversity, stability and economics. We can all contribute to this minimum ethic, particularly through positive action – letting the fly go. Such a gesture is also an ethical templet for our children.

9.1.2 Value-perspectives on individuals versus species

Our value system accords rights to human individuals, not to our species. Giving rights to other species is therefore difficult to defend given our normative viewpoint. Concern for highly threatened **species**, not the last remaining **individuals**, is an emotional response through fear of irretrievable loss. **Individuals** of a threatened species could be considered to have no more rights than those of an abundant species, given equal levels of sentience. A moral concern and helping hand starts to become justified for individuals of a threatened species when those individuals are subject to escalating physiological stress as external conditions increasingly militate against their survival.

Even qualitative decisions based on comparative values of different species are difficult to justify. Given the choice between two species, we instinctively opt for the one with apparently higher sentience, the elephant over the fly. We also opt for aesthetics, size and flamboyance over the converse, even if the least glamorous is taxonomically remote and genetically unusual. We prefer the threatened swallowtail butterfly to the threatened zorapteran. A zorapteran individual is hardly even a feature in the whole human thought process at any one time (Figure 9.1).

The moral dilemmas of deep, bioethical values (Callicott, 1986; Naess, 1989) versus the so-called shallow, anthropocentric value of insect variety are not easily reconcilable given the principle of natural selection. Considering the gruesome activities of parasitoids upon their hosts, nature has not acknowledged the safety of individuals, even from other individuals of the same species, whether cannibalism or adelphoparasitism. Nor has nature made any guarantees for species, which evolve or die out. These issues are big issues in conservation biology, as insects are so numerous as individuals and as species, and they play such an important role in ecosystem function.

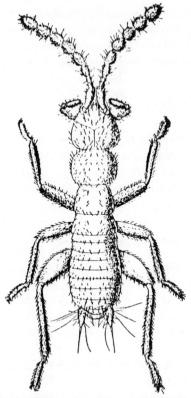

Figure 9.1 An individual of *Zorotypus guinensis* (order: Zoraptera) – taxonomically remote, and remote from human thought. Who will care for such an individual, a species, an order? (From Richards and Davies, 1977. After Silvestri, 1913.)

Before tool-making man, other events, both biotic and abiotic, caused the extinctions. We are causing, by our own competitive dominance, widespread stress to many individuals and causing many species extinctions. And we are fully conscious that we are doing it. And much of the conscious effort to preserve species is for our own welfare and benefit, which is shallowly anthropocentric.

With the theoretical exception of a species conservation procedure threatening the survival of another species, species preservation is not necessarily shallow. Species conservation may be considered shallow if simply a form of resource management. Although the individual and not the species is the raw stuff of evolution, it is the genome that is the continuum. From that genome arises the individuals, the almost perfect symbiosis between gene and the body manifested by it – the phenotype. From the genome comes the future generations of individuals. Conservation, therefore, should consider preservation of the genotype–phenotype symbiosis. This approach, takes in population dynamics, and marries species conservation with the Lockwood ethic. This is surely not shallow or singularly anthropocentric.

9.2 INSECT UTILITARIAN VALUE

9.2.1 From individuals and species to ecosystems and landscapes

Norton (1986) has pointed out that there are utilitarian reasons of varying degrees of persuasiveness for humans to preserve other species. Commercial use is uppermost, even though our present economic system may not necessarily be an appropriate ethical underpinning to a future sustainable world (Schumacher, 1973). Given the vast number of species, preservation of a species is not convincing in commercial terms because the probability that any given species will be commercially useful is relatively low. Also, until a particular species is shown to be useful, there is no way to place a quantified commercial value on potential benefits, or to decide whether a given species is uniquely able to provide them. So, in hard commercial terms, general preservation of all species has high risk returns. Moving to a larger scale of measurement, and considering interacting individuals and species, may have more utilitarian value.

Humans derive many important services from communities, ecosystems and landscapes. These include purifying water, hydrological and atmospheric stabilization, as well as many services from insects such as natural control of pests, genetic resources for biological control, pollination of natural plants and crops and decomposition. Ecosystems also supply the raw material for rearing of insect livestock and deadstock.

Preservation of varying and overlapping ecosystems has a sound base ecologically, pragmatically and aesthetically. Not only may the ecosystems and the landscape be attractive to the human eye, but the insect species and other biota are an intrinsic part of it. As ecosystems are composed of individuals and classes of similar individuals (species) interacting among themselves and with their physical environment, these smaller open-system units are the ultimate providers of those benefits. Such services are usually non-commercial and not easily priced, yet the price of providing and maintaining them singly and artificially would be high in both money and energy (Norton, 1986).

The cogent utilitarian argument for species preservation comes from risks of rapid ecosystem change as plant and animal species are gradually eliminated. The permutation of interactions increases proportionately with the increase in number of species. Working in reverse, when does the point come that an ecosystem drastically changes as species, particularly key species, are removed?

Because of the vastness of the number of interactions, as well as number of species, there is difficulty in deciding on the usefulness of cost–benefit analysis as a method of determining which species preservation efforts are socially beneficial. As one moves away from obvious single-species commercial values (as in biocontrol), cost–benefit analyses become increasingly difficult to propose and defend. Norton (1986) points out that, as broader categories of value are included in the analysis, it becomes increasingly difficult to advance non-controversial dollar-value assignments.

Aesthetic values and various other services of ecosystems are particularly difficult to quantify, although less so for landscapes. Even if one could assign replacement costs to the various services, there is a lack of knowledge of the comparative contributions of species to the provision of services. It is therefore unclear how to apportion benefits to the individual species that compose a system. Also, there is the arbitrariness of quantifying risks associated with ecosystem and landscape degradation.

There is more merit in a qualitative goal such as reducing known risks to species and ecosystems than attempting to precisely quantify all risks. Quantification of risks requires a time frame for computing costs, which is an untenable approach. Since species, once lost, cannot be recreated, a decision to allow a species to become extinct is irreversible. Any values accruing to humans through indefinite time would be lost. But quantitative analysis implies limited time, and cannot therefore represent the fullest long-term value of a species. The value of a species generally increases the longer the time of the benefits. It then follows that decisions to preserve a species become more valid as one places greater weight on the obligations of present generations to protect the future (Norton, 1986).

9.2.2 Insect services

Use of insects for biological control, pollination and honey production has varying benefits to individuals and to the species. It is the insects' services or their products that are being traded. With insect biocontrol, there is the double-edged ethical argument, in that one set of individuals of one species is benefiting at the expense of another. In weed biocontrol, the insect is benefiting at the expense of the plant.

With pollination, all species benefit, with the understanding that a crop system may be taking the place of a natural ecosystem, or that a weed is being pollinated that is invading an otherwise natural landscape. With honey production, the honeybee is conserved and populations promoted, but nevertheless some individuals are killed during hive making.

These services might be viewed as simply exaggerating natural activities and processes, and cannot, in general, be said to be overexploiting nature. One exception is that some insect biocontrol projects apparently have caused some non-target insect species to become extinct.

9.2.3 Insect farming

Invoking insect services is akin to farming any other type of animal for services, such as sheep for wool. Significant is that the insects probably suffer less than some other animals when providing these services – consider battery hens producing eggs (Dawkins, 1989). But not all forms of insect utility are so free of insect hurt.

With insect farming as with other types of farming, there is inevitably killing of the insect. There is also a trend towards artificial selection away from the wild type for better production. In centuries of silviculture, *Bombyx mori* has been selected for finer quality silk and greater production levels. Also, it is no longer known as a wild species.

Interestingly, there are some other parallels with more familiar farming principles. The demand for insect specimens by collectors, and in particular the demand for unblemished ones, stimulated the development of butterfly farms, particularly in Britain. Today, butterfly farming is also undertaken in the tropics, especially South-East Asia and Papua New Guinea, providing employment for the local human inhabitants. This takes the pressure off the wild butterfly populations, especially the birdwings.

Over recent years in technologically advanced countries, there has been greater human appreciation of the aesthetics of nature as leisure time has increased. A market has developed where there is a demand to see living insects.

In 1989, Britain alone had 38 commercial butterfly houses in operation displaying living butterflies to the public (Morris *et al.*, 1991). Although these houses have displayed over 300 species of Lepidoptera, only about two dozen or so are the core attraction (Collins, 1987). Some butterfly houses show up to 70 species at one time, and about 150 over the summer season. About 500 000 individual butterflies were used in 1986, one-third being bred on site and two-thirds brought in from dealers in Britain and various tropical countries. There is no evidence that threatened species are used, and virtually all the material is captive bred (Morris *et al.*, 1991).

The deadstock trade runs into tens of millions of dollars annually. Most of this trade is for the decorative ornament and display trade, using large numbers of mostly bred, common butterflies. The much smaller specialist trade is aimed at serious collectors who demand good-quality individuals of rare species, US $1500 having been paid for a male hybrid birdwing butterfly *Ornithoptera 'allottei'* in Paris in 1966. Many of these specialist trade species are rare and threatened. In several countries collecting or dealing in such species is illegal, while for CITES-listed species (Convention on International Trade in Endangered Species of Wild Fauna and Flora), international trade is banned or restricted.

The ethical arguments for and against these activities parallel those on use of other, more sentient, animals. The deadstock ornament and display trade is wholesale slaughter for luxury (yet low-value) items. They are also items that decrease, not increase, in commercial value. They are also readily destroyed by invasions from fungi and, ironically, from insects such as dermestid beetles. They also deteriorate with prolonged exposure to sunlight. On ethical grounds, there is no moral justification for use of insects in this way.

The specialist scientific trade has quite a different complexion. Very few individuals are destroyed and the carcasses of the dead specimens are generally well preserved, often eventually being deposited in museums. As well as being of scientific value as reference specimens for various subjects from systematics to conservation biology, they are also depositories of genetic information for DNA technology.

The risk of the deadstock trade is that fanaticism may overrule reason, with unscrupulous collecting of highly localized, low-density species taken from the wild. This does not seem to be a general threat, and nowhere is the threat as great as that of habitat destruction. Nevertheless, it requires alert monitoring (New and Collins, 1991), as recently seen in the case of extremely rare and flightless South African *Colophon* spp. stag beetles, that only live on mountain peaks, and which have appeared on the market for as much as US$ 11 000 each.

9.3 VALUE OF THE LANDSCAPE

9.3.1 Land ethic and the entomologist

When we think of the landscape, we tend to conjure up impressions of the topography, landscape pattern and the arrangement of trees. What we do not see at a glance is the interactiveness of all the organisms, plant and animal, large and small. Over the past few decades views have changed and we now consider not only the interactiveness of organisms that so fascinated Elton (1927, 1958), but also that the land is a community to be respected, a biotic mechanism and a fountain of energy flowing through a circuit of soils, water, plants and animals and back to the soil.

Leopold's (1949) sensitivity to the adverse effects of self-interest of human individuals, corporations and governments has increasingly overridden the obligations to the land, which is the inheritance of future generations. In other words, a land ethic implies respect by human individuals for their fellow humans and also respect for the biotic community (Rolston, 1986, 1988; Taylor, 1986; Potter, 1988).

As biotopes, ecosystems and the landscape are the determinants of insect survival, a land ethic is fundamental to insect conservation. This leads again to the importance of supplying man's needs sustainably, as well as those of other organisms, including insects.

Entomologists are caught in a dilemma. Many are employed to improve crops, whether oil palm, potatoes or pines. Attention is focused on the crop and its economic pest. The emphasis is on economic maximization of the crop with rarely any regard for the neighbouring landscape elements, except as refugia for the pests' natural enemies. This is a task, an economic task, which may or may not consider the land ethic depending on the type of impact that the control measures may have. Biological control of, say, a specific forest pest may avoid the use of an insecticide that carries the risk of permeating the soil. In contrast, the biocontrol agent essentially must be highly specific to the pest, if it too is to be ethically acceptable. Translocations leading to rampant population expansion are not in keeping with the land ethic.

The view of such an applied entomologist may not consider the landscape in general, and this is the area where entomologists need to play a far greater role in achieving sustainability both within the crop (e.g. allowing natural vegetation to thrive as far as possible) and between the crops. Sustainability can also be promoted despite landscape fragmentation, by providing refugia and movement corridors in keeping with the wild-community ethic. Corridors should be of native plant species rather than introduced ones wherever possible.

Tampering with nature must be done with sensitivity and planning. For insects, this means planning for future human **and** insect populations, as well as for immediate control of pests. Working with nature, rather than blatant dominance, is the approach. Entomologists must be aware of the land ethic, as environmentalists must be aware of the importance of insects in ecosystems (Rolston, 1988), to put the plan into action. The entomologist is naturally honing in on one, or at most a few, species of economic significance. As it is not possible to give individual attention to each of the myriad of species outside and even inside the crop, it is vital to be aware of the landscape pattern as a catch-all for the other, often nameless, species and their interactions. Conservation and sustainability recognize the target pest species in one cultivated patch, while at the same time encouraging maintenance of as many as possible, as large as possible, natural patches.

9.3.2 Priority systems and avoiding triage

Species are being lost to extinction probably at a rate at least 1000 times faster than any time in the past. This is the result of our collective human impact. It is the human conscience, both for intrinsic value and for man's benefit, that is now deciding to put the brakes on pushing species into the abyss of forever gone.

To save species requires painful decisions to be made as to which species deserve a place on the ark. One approach to prioritizing for preservation of species is the system of triage. Triage was the French policy of sorting wartime casualties into three categories for medical attention: those with wounds so

severe that treatment would not avert death, those with wounds not sufficiently severe to require immediate attention, and those in between with
serious but treatable wounds (Norton, 1987).

Given the limited funds and time, is prioritizing fruitful? Once the problem
of threatened species is looked at more broadly, as a problem of protecting
overall biological diversity, rather than a problem of saving individual
species, it can be restated in a different way. How ought insufficient funds
and efforts best be spent to meet threats to biological diversity? This approach
avoids the triage solution that sorts species into those that will receive
attention and those that will not. Reformulating the problem into a saving
of the total diversity of areas does not make the problem go away, but it does
give it more meaningful ecological and evolutionary perspective. Having a
broad range of species available in each area as potential colonizers and
community components strengthens the forces that lead to diversity in
successional communities (Norton, 1987).

This conservation of ecosystem dynamics is particularly pertinent to insect
conservation biology. Insects show enormous variation, at different life
stages, between sexes, between morphs and between species. They are present
in vast numbers as individuals and as species. They also often show considerable demographic volatility, and are highly interactive in communities at
various trophic levels. Prioritizing then becomes a futile exercise given the
magnitude of the overall problem.

This calls into question the ethical value of *Red Data Books*. Although not
stated, there is an implicit suggestion that those species closest to the Extinct
and Endangered categories require greatest input of funds for rediscovery,
research and practical management. This is a particularly tenuous approach
with insects, when really we know so little of their distributional and
population viability status. RDBs are good highlighters, but to be used
alongside selection of landscapes for species and ecosystem variety and
connectiveness.

9.3.3 Ecosystem and landscape value

As insects are unreservedly major components of almost all terrestrial ecosystems, we cannot close our eyes to ecosystem and landscape value. A
difficulty with valuing ecosystems is that they have no subjectivity, no felt
experiences. They are communities mostly of living objects (non-emotively
valued) and of living subjects (emotively valued). Duties can **concern** ecosystems but must be **to** subjects (Rolston, 1988). Ecosystems and landscapes are
the vehicles for individuals, populations, patterns and processes. This scale
of measurement (at the ecosystem and landscape level) is an interactive and
integrative platform for the ecological objects and subjects in the evolutionary
story. But the platform does not have well-defined edges. There is inflow and

outflow of materials and individuals. The quantity of interactiveness and continual change in interaction strength in an ecosystem is so vast that it is scarcely measurable. Hence so many of the potential ecological paradigms of this century have had to be recategorized as generalized hypotheses, supported or rejected by the various taxonomically or geographically localized investigations.

Concern for ecosystems is therefore of great importance for all its individuals, species and interactions. We cannot comfortably hug a bug **species** or embrace an ecosystem. We can, though, see and touch an individual insect. We may, on rare occasions, be able to see a whole population. Apart from an unusually small group of individuals making up the sole last remnant of a species, we cannot **see a species**. Nor can we see a habitat, just as we cannot see a niche. Nor can we see, touch or smell an entire ecosystem. We only catch glimpses of an ecosystem at one time in one place. Our senses filter our perceptions, which modify our judgements.

Habitat structure, a plantscape, a biotope and a landscape are principally structural entities by definition. Being more tangible, they are more comprehensible than other ecological concepts. Nevertheless, they embody interactions, and the structure we see is transient. We can relate to them and fairly readily call for their conservation.

A biotope or a landscape does not feel pain and nor is it sentient. With this in mind, it is difficult for us to have a truly intrinsic, subjective concern for these larger scale, interactive, open systems. But we see our impact upon these dynamic structures. Our conscience, coupled with recognition of our power of impact, steers us towards rational management for intrinsic and pragmatic worth of these biotic systems.

Although it hurts to see toppled ecosystems, the recognition of the damage is subjective. The forester may see great value and meaningful achievement as he or she looks at a neatly arranged patch of planted trees. To the wilderness lover, it is an ugly sight and a wrongful act.

Within the framework of an ethical conscience, but with a practical awareness that food and materials have to be teased out of nature's fabric, concern for individuals, populations, species, assemblages, communities and ecosystems comes about by protection, appropriate management and awareness of all interacting life forms at the level of the landscape. Visually and photographically, we can, to a large degree, measure many landscape patterns and processes. We can also manage at this coarse scale with a fairly good sense of value and feeling that many smaller scale entities and processes will continue under the umbrella of landscape management. This scaling up continues beyond the landscape. All the networks of landscapes make up the biogeosphere and the atmosphere, the interacting components of Gaia. In the final analysis, care for insects, care for life, comes about from care for the planet.

In holistic terms, insects are major components of terrestrial ecosystems

indirectly involved in Gaia maintenance by being the major faunistic interactors with plants in many ways. They nibble, penetrate and pollinate
Lovelock's (1988) 'daisies'.

9.4 SUMMARY

Human individuals generally have little esteem for the individual insect.
Nevertheless, there is evidence for insect sentience. This means that we should
refrain from actions which may be unreasonably expected to kill or cause
non-trivial pain in insects when avoiding these actions, has no, or only trivial,
costs to our own welfare.

Our value system accords rights to human individuals, not to our species,
making moral extensions to other species difficult. Evolutionary lines have
ruthlessly removed the majority of individuals and species over the millenia,
and we are today speeding up the extirpations. Although the **species** does not
feel pain, the individual and the species are nevertheless inextricably linked
by the genotype–phenotype symbiosis. This makes ethics regarding the
individual closely linked with that of the species.

In utilitarian terms, humans derive many benefits from ecosystems, and
many direct and indirect benefits from their insect components. Preservation
of ecosystems in the landscape maintains the vast array of interactions as well
as all the individuals and species, most of which we cannot even catalogue,
let alone understand. There is merit in reducing known risks to ecosystems
for maintaining biodiversity, of which insects form a major part.

Insects provide many valuable services, the employment of which are
generally ethically supported. Insect farming for aesthetics is also generally
acceptable, although the wholesale slaughter of millions of farmed or wild-
caught butterflies and other large insects for garish ornaments has little
justification. The specialist deadstock trade eliminates few individuals and
has great scientific merit, making it more morally acceptable.

The landscape ethic implies avoidance of self-interest exploitation of the
land. Beside wilderness areas, there needs to be ecosystems that are utilized
at most on a sustainable basis. Entomologists need to be aware of the land
ethic, as environmental ethicists need to be aware of the major ecological role
that insects play.

Individual and species protection is pragmatically and ethically approached by biodiversity conservation, which, in turn, implies conservation
of ecosystems and landscapes. The scale of measurement of the landscape is
conceptually and managerially viable and valuable. Landscapes can be
physically seen, touched and measured. Management decisions can be made
which cover all the possible interactions between individuals and species
within biotopes and ecosystems making up that landscape. Concern for
landscape is concern for all life and its processes. This landscape approach

also creates an awareness and an ethic that all landscapes are linked by biogeocycles. Insects are major interactors in the whole terrestrial biosphere and are therefore party to Gaia maintenance.

Part Four
Positive Action

-10

Insects, the landscape and evaluation

The species...

AD 1857. Hot summer. The two Clouded Yellow
Butterflies plentiful... Camberwell Beauties and
Convolvulus Hawks. The Rocky Mountain Locust
devastates in America. Oleander Hawk taken at
Brighton. The Large Copper Butterfly has disappeared
from our fens previous to this date.

A.H. Swinton (c. 1880)
(The Rocky Mountain
Locust is now extinct).

Without taxonomy to give shape to the bricks, and
systematics to tell us how to put them together, the
house of biological science is a meaningless jumble.

Robert M. May

The landscape....

Within the histories of species, individuals are
perpetually perishing, but species are prolonged until
no longer fit in their environments – whereupon they
evolve into something else or go extinct and are
replaced. Beauty does not require permanence. Within
landscapes there is ugliness in the detail, but at the
systematic level, at the scope of the dynamic scene,
softened by perspective from a distance, there is
sublime beauty .... Great beauty, like great music, is
often a minor key.

Holmes Rolston, III (1988)

10.1 EVALUATION TO ACTION

10.1.1 Significance of landscape conservation

Whether from an ethical or pragmatic viewpoint, insect individuals, populations, species, and all the interactions, are protected when the landscape is conserved. The landscape is sufficiently large scale and of a fairly well-definable structural composition for it to be a good umbrella for preservation of smaller scale units and processes (Figure 10.1). This includes conservation of insects and their habitats. This approach does not ignore the fact that the landscape is a changing system. Although the changing species profiles of landscapes are inherent to their dynamics, the species and species assemblages are a reflection of past events (Burel, 1992).

This umbrella approach recognizes larger scale regional and historical influences, which are the backcloth for local and contemporary interactions

Regional Processes
(historical and contemporary)
↓

	Landscape		
	Structure (physical three-dimensionality, including fractals)	Composition (heterogeneity, i.e. differences between units)	Interactions (within and between individuals, populations, species and landscape elements)
Decreasing scale	1. Topography	1. Matrices, patches and corridors	1. Ecosystems
	2. Matrices, patches and corridors (i.e. landscape elements)	2. Habitats (i.e. the place where an insect lives relative to its behavioural ecology and life-history style)	2. Communities
	3. Biotope	3. Microhabitats	3. Metapopulations, populations
	4. Plantscape		4. Interspecific interactions
	5. Microsite		5. Intraspecific interactions

Figure 10.1 The landscape is a useful umbrella option for evaluation for insect conservation. The landscape level embraces most of the various local influences on a relatively mobile but small terrestrial animal like an insect. Landscape structure, composition and interactions are influenced by historical and contemporary regional processes, such as climate and changes in species range . Structure, composition and interactions are all interrelated and linked. This is simply a schematic representation. The value of the landscape approach is shown by the fact that the landscape elements (matrices, patches and corridors) can be listed under both 'structure' and 'composition'. The decreasing scale is approximate, with occasional inversions (e.g. some populations may be larger than other metapopulations).

and processes. For example, changes in regional climate may also affect the plantscape, which in turn influences the outcome of competitive interactions between species (Figure 10.1). Nor does the landscape approach ignore local, natural population variations. These variations include local population extinctions not directly mediated by man. Such local extinctions can occur in various types of landscape from temperate lands through to the tropics (Dempster, 1975; Murphy, Freas and Weiss, 1990; Wolda, 1992). Population and species extinctions may be directly (e.g. cutting down of trees) or indirectly (e.g. leaving fragments which are too small and too far apart to support viable populations of insects) caused by man modifying the land-scape. Both influences may cause extinctions, as in the case of butterfly populations in the Netherlands (van Swaay, 1990).

The significance of the interaction between different scale processes, including the influence of man's modification of the landscape, is seen in the interesting case history of the extinction of the Rocky Mountain grasshopper (*Melanoplus spretus*) (Lockwood and De Brey, 1990) (Figure 10.2). This grasshopper was once so widespread and abundant that it was a serious agricultural pest in the western United States and Canada before the year 1880. In 1857, 1864, 1867, 1869, 1872 and 1874 the grasshopper was described as devastating the Western United States (Swinton, *c.* 1880). There then was an enormous widespread population crash in the early 1880s due apparently to a combination of natural factors including predation and particularly disease. After the crash, the grasshopper's range was reduced and fragmented, exacerbated by loss of rangeland to agriculture. In the late 1800s, oviposition and early nymphal development occurred almost exclusively in association with riparian vegetation. Intensive modification to these remnant patches through tillage, irrigation, loss of the beaver and introduction of cattle, plants and birds, were instrumental in causing the final extinction of the species.

These points emphasize the importance of scale of measurement and size of landscape elements in maintenance of an intact interactive biota. In terms of scale, very different pictures of patterns and processes emerge depending on whether quadrats are 1 m × 1 m, 10 m × 10 m, 100 m × 100 m, 1000 m × 1000 m, etc. In addition, time scale is important. The interactions and processes, say between insects and plants, or the insect herbivores and their parasitoids, vary over time. The implications for conservation biology are that not only may size of an area be important for maintenance of local populations, but factors such as toposcape, plantscape and connectivity may also be important. A particular pattern of plant assemblages across a certain topographic landscape and the presence of movement corridors are likely to be crucial for survival.

There are two interrelated approaches to viewing insects within the context of nature conservation. Landscapes may be conserved to protect the

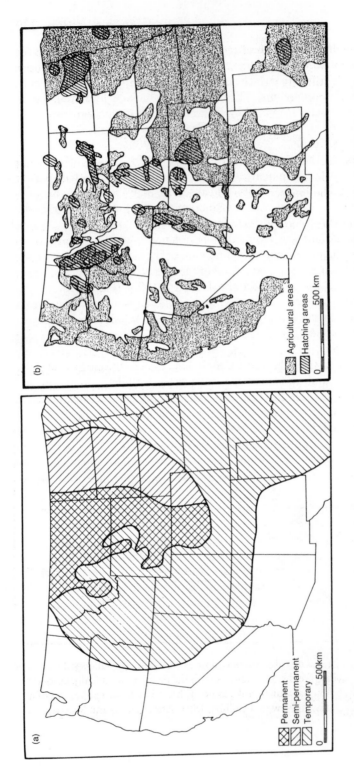

Figure 10.2 (a) Distribution around the middle of the last century of the Rocky Mountain grasshopper (*Melanoplus spretus*) in the western United States. The depicted area is most relevant to the ecological events during the extinction. The full distribution of this species extended into Canada and the central United States. (b) After a natural population crash, the cultivated areas for maize, wheat or hay continued to fragment and suppress oviposition, and hatching areas were under cultivation (right). The species was last seen alive in 1902. (From Lockwood and De Brey, 1990.)

insects, along with other biota and geology. Alternatively, insects themselves may be used in the evaluation process. Surveying the component insect species may pinpoint scientifically valuable sites, which, once identified, also conserved other biotic components.

10.1.2 Landscape conservation: the global perspective

Around the world, landscapes are being modified both extensively and intensively. It is valuable scientifically to identify those areas of the world where modifying the landscape may have most impact on faunas of great value.

On the global scale, as well as identifying world 'hotspots' and world 'wilderness areas', which cut across political boundaries, specific countries have been identified as 'megadiversity countries' (section 3.1). The 'critical faunas' approach of Ackery and Vane-Wright (1984) has also been useful in identifying those countries which have genetically valuable faunas that are in need of protection at the national and international level. This has been done for the world's swallowtail butterflies by Collins and Morris (1985), who carried out a distributional analysis of the world's swallowtails, recognizing 51 countries with 'critical swallowtail faunas' (Figure 10.3). Forty-three of these countries support endemic species, and they can be ranked according to the number of these endemics. Indonesia, with 53 endemic taxa out of a

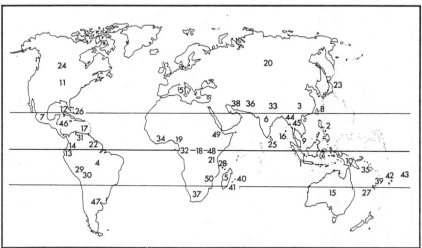

Figure 10.3 Ranking and distribution of countries with critical swallowtail butterfly faunas. These are the 51 countries with the most taxonomically unique and rare species. Indonesia, which ranks as no. 1, has 53 endemic species out of a total of 121 species. (From New and Collins, 1991. After Collins and Morris, 1985.)

Table 10.1 Critical swallowtail faunas (CSFs) (after Collins and Morris, 1985) and critical tiger beetle faunas (CTBFs) (after Pearson and Cassola, 1992) for the first 10 countries for each insect assemblage

	CSF					CTBF			
A	B	C	D	E	A	B	C	D	E
1. Indonesia (121)	53	68	121	121	1. Madagascar (176)	174	2	176	176
2. Philippines (49)	21	4	25	146	2. Indonesia (217)	103	114	217	393
3. China (104)	15	61	76	222	3. Brazil (184)	97	87	184	577
4. Brazil (74)	11	63	74	296	4. India (193)	82	88	170	747
5. Madagascar (13)	10	3	13	309	5. Philippines (94)	74	2	76	823
6. India (77)	6	8	14	323	6. Australia (81)	72	4	76	899
7. Mexico (52)	5	37	42	365	7. Mexico (116)	57	58	115	1 014
8. Taiwan (32)	5	0	5	370	8. USA (111)	45	24	69	1 083
9. Malaysia (56)	4	1	5	375	9. PNG (72)	45	1	46	1 129
10. PNG (37)	4	8	12	387	10. RSA (94)	40	52	92	1 221

A = Country (total number of species), B = number of endemic species, C = non-endemic species not occurring in previous countries, D = newly acceptable species, E = cumulative species total.
PNG = Papua New Guinea; RSA = Republic of South Africa

total of 121 species, ranks as no. 1. The Philippines has 21 out of 49 (no. 2), China 15 out of 104 (no. 3), Brazil 11 out of 74 (no. 4) and Madagascar 10 out of 13 (no. 5). These first five countries support nearly 54% (309) of the world's swallowtail species. If the next five countries (India, Mexico, Taiwan, Malaysia and Papua New Guinea) are added, the figure rises to 68% (387). Thus, if the swallowtails of the first 10 countries are conserved, this would account for 387 out of a total of 573 species. The swallowtail critical faunas analysis has been translated into an IUCN-ratified Action Plan (New and Collins, 1991). An Action Plan, as the name implies, is a positive strategy for conserving the subject insects, based on the best available evidence.

It is now timely to use this first insect conservation Action Plan as a yardstick for determining other critical faunas. Pearson and Cassola (1992) have done this for critical tiger beetle (Cicindelidae) faunas. The critical faunas lists in Table 10.1 show that 387 out of the world total of 573 swallowtail species (i.e. 68%) occur in just 10 countries, while 1221 out of 2028 tiger beetle species (i.e. 60%) occur in their first 10 countries. There are some implicit assumptions in these figures e.g. (1) that all the species have been discovered, (2) that there is taxonomic stability, (3) that there are no extinctions, (4) this is a single time frame, etc. [see Collins and Morris (1985) and Pearson and Cassola (1992) for details]. Nevertheless, the overall trends do indicate that some countries are extremely important in terms of numbers of species, especially endemic ones. Critical faunas of swallowtails and tiger beetles occurring in the two series of first 10 countries are in Indonesia, Philippines, Brazil, Madagascar, India, Mexico and Papua New Guinea. Seven out of 10 countries are common to both lists. Naturally these positions can shift with greater knowledge, but it does give a picture of which areas/countries of the world are holding the genetically most scientifically valuable insect faunas.

Such large-scale surveys depend on a vast background of taxonomic knowledge. This of course is not yet possible for many other groups of insects. Even on a smaller scale of landscape and its component interacting ecosystems, such a critical faunas approach could prove valuable. Some insect groups are well suited to this approach as they have been shaped to both larger and longer scale processes and yet are also responsive to local biotope change. The cicindelid beetles have shown themselves to be particularly valuable, because as well as having these features they are morphologically distinctive and present no taxonomic impediment (Pearson, 1992; Pearson and Cassola, 1992).

One cautionary note with the critical faunas approach is that it views biodiversity almost as a museum collection of evolutionary dead-enders. Conservation biology has a futuristic element as well as a preservationist one. It is as important to conserve species dynamo areas [e.g. the central African savanna areas for acridids (Jago, 1973)] as it is to conserve areas of genetic uniqueness. The dynamo areas are the evolutionary stepping stones to future biodiversity. The centres of radiation could also be linked by large corridors (Erwin, 1991). It is still possible to conceive and plan these in the wilderness areas of the world (section 11.1 and Figure 11.5).

10.1.3 Landscape conservation: costs and constraints

Any conservation evaluation programme must consider costs and constraints. As nature conservation can involve vast geographical areas and deals with a wide variety of approaches and a great variety of life, especially among insects, cost-effectiveness is vital. Time is short for conserving as much of the world's biota as possible. The most cost-effective approach to preserving as many insect species as possible is to conserve a variety of landscapes (section 10.1.6).

An important first step is to be familiar with theoretical ecological studies pertaining to the scale of investigation. Patterns, composition (biotope heterogeneity) and processes vary considerably from one scale to another, while smaller scale measures may overlap each other and are nested in the larger scale measures (section 4.1). At the finer scale, a knowledge of genetic bottlenecks and the 'single large or several small patches or reserves' debate is useful (section 5.2). On the coarser scale, the value of movement corridors may be important (section 5.4). An awareness of possible future global impacts overlays all scales of investigation. The value of these background ecological tenets and proposed principles is that they are useful for formulating hypotheses that can statistically be accepted or refuted. Conservation biology is then in step with other areas of scientific investigation, giving it scientific credibility.

10.1.4 Landscape conservation: importance of setting of goals

Conservation evaluation must have clearly defined goals. This is necessary as the field is vast, biotic communities are immensely complex, land is finite and time for proactive management against species loss is short. Goals may be, for example, to identify areas of particularly high endemism. Or they may be to measure continuity (i.e. insect traffic flow) between landscape elements. In contrast the aim may be to find an optimum means of integrating local human needs relative to the biotic value of an area. There are many possible goals, which are by no means mutually exclusive.

10.1.5 Landscape conservation: some approaches to diversity measurement

Data and information need to be gathered and analysed relative to the goals set. Where, say, the aim is to locate areas of high endemism, then collecting presence or absence data (i.e. measurement at the nominal level) may be the most direct means. This in itself may be a vast task, especially in tropical and savanna areas. It may be more expedient to hone in on a particular insect taxon. Presence/absence data measure species richness. As insects are the numerical majority, this is a measure of biodiversity, especially when coupled with measures of plant richness.

Simply assessing the numbers of species present, especially with un-named morphospecies, makes no value judgements on the taxonomic status or genetic uniqueness of the fauna. Once species are named, they can be related to earlier studies, and levels of endemism assessed.

Another useful measure is diversity measurement, which is the centre of biodiversity studies. This is discussed in greater detail later in this chapter. Suffice to say here that it is measurement at the interval level (i.e. measuring absolute levels of populations of different species). Such assessments at any one time are only one small window on reality. Insects often show great variability in population levels, and characteristically often have highly varying spatiotemporal population patterns.

Concepts of rarity (section 7.1) are also closely coupled with diversity measures. Not only does the relative number of rare species influence the diversity measure, but the number of species identify a site as perhaps appropriate for conservation of these rare species. This assumes that rare species conservation is the goal. In turn, rarity may or may not be associated with localness of distribution and taxonomic uniqueness. It is rare, localized and taxonomically remote species that are the ones often under most threat and are in the *Red List* (IUCN, 1990) or in one or more *Red Data Books*.

A general method for evaluating insect rarity is to assess the presence or absence of species in relation to a grid. This is illustrated in Figure 10.4, where the outline could be a county, region, state or country and the grid lines may be at 1-, 2-, 10- or 100-km intervals. Species A, B, and C in Figure 10.4 occur, respectively, in 3, 9 and 50 of the 150 grid squares, and their frequencies of occurrence are 0.02, 0.06 and 0.33 respectively. The inverses of these are 50, 16.7 and 3, which provide reasonable, geographically based, indices of rarity (Usher, 1986b).

10.1.6 Landscape conservation: 'naturalness', 'typicalness' and 'representativeness'

One approach to landscape evaluation is to find sites that have suffered little human disturbance, i.e. which are high in 'naturalness'. Sometimes this is difficult to determine, if determinable at all, where man has influenced the landscape for centuries. Surveys of the insect and plant species require a good taxonomic knowledge to determine naturalness. As landscape modification

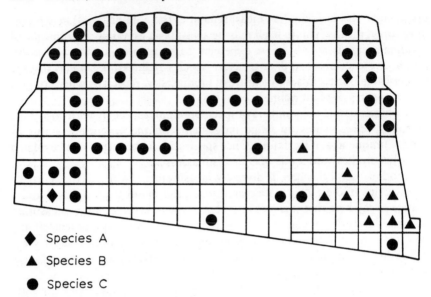

♦ Species A

▲ Species B

● Species C

Figure 10.4 The distribution of three species in relation to a grid covering a hypothetical area. Such a grid could cover a large area such as a country or state, or a smaller area, such as a local district or landscape. The grid is accordingly scaled up or down. Note that there are difficulties in defining 'squares' at the margins: generally a marginal area has been counted as a separate grid square if it occupies half or more of the area of a complete square. Thus the marginal squares have areas that range from 0.5–1.5 times a complete square. See text for discussion of species A, B and C. (From Usher, 1986b.)

and disturbance tend to reduce species richness, naturalness and high diversity tend to be strongly correlated (Usher, 1986b).

Given a large land area, it may be useful to identify sites that are typical in terms of their landscape and species. Conservation for 'typicalness' is the converse of conservation of rarity, although both approaches have great conservation merit. Determination of which is the most typical site involves listing species biotopes and ecosystems in a geographical area. For insects, recognition and classification of the plant species and plantscape is an important prerequisite. This is particularly so as high local plant endemism is usually associated with high local insect endemism, and high plant species richness **plus** complex plantscape architecture leads to high insect species richness.

Another useful concept is 'representativeness', which is a relative term ranking sites in terms of the ecosystems occurring within them (Austin and Margules, 1986; Usher, 1986b). It is often a preliminary step in any evaluation and enables a decision to be made on which site or sites best have typical examples of species, biotopes, communities and ecosystems. 'Representativeness' is therefore a comparative term, whereas 'typicalness' can be more localized and applied to an individual site (Usher, 1986b). Representativeness should be based on biologically relevant attributes rather than simply climate, lithology or any other environmental variable. Representativeness uses a hierarchical classification of land units which allows the level of heterogeneity, at which the representative samples are to be taken, to be clearly specified. This is a particularly important approach for insects, as so many species are undescribed, and for areas where the proportion of undescribed to described species is high. Representativeness should contain biota which represent the range of variation found within a particular land class or system. As Austin and Margules (1986) point out, the approach is quantitative and the evaluation covers a range of scales from individual sites to the global scale. In reality, the flora forms the thrust of the approach, but there is no reason for not overlying the same criteria for insects, even if they are relegated to nameless realms of morphospecies 1 to *i*. Morphospecies initially can be recognized by their 'jizz' (Disney, 1986) and allocated full taxonomic descriptions later. This does not preclude recognizing that something is rare, localized and valuable, even if temporarily nameless.

Further discussions on these and other criteria used in general nature conservation evaluation are given by Spellerberg (1981), Usher (1986a), Saunders *et al.* (1987) and Margules and Austin (1991).

10.1.7 Landscape conservation: the management factor

Evaluation, especially with regard to insect conservation, is not necessarily only for natural ecosystems. With so many modified landscapes across the world, evaluation also involves assessing modified landscapes for their biotic value. The value may be cast in terms of sustainable agricultural value for natural predators of insect and mite pests. The evaluation may be for repatterning of the landscape to promote population levels of certain insects of intrinsic value (as with conservation headlands for butterflies) or to encourage more improved natural enemy activity. Conservation of insects within the agricultural landscape is most important, bearing in mind the need to reconcile agricultural output with conservation of species. In assessing farmland in Britain, Cobham and Rowe (1986) found that species richness, especially of plants (hence of insects), was particularly useful in determining the conservation value of agricultural sites.

Evaluation of remnant patches in an agricultural matrix must take into account the impact of that matrix on the patches. Webb (1989) has argued, based on data from beetles and spiders in southern English heathland remnant patches, that the evaluation should include criteria based on the physical and

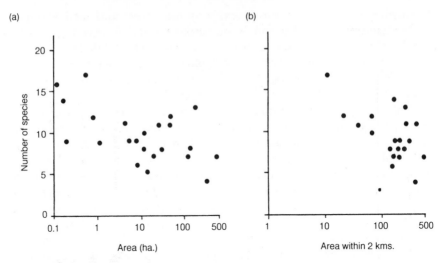

Figure 10.5. (a) Decrease in point species richness of phytophagous Coleoptera within 22 small English heathland remnant patches as the patch area increases. (b) Decrease in species richness in small remnant patches with an increase in the area of large heathland areas within 2 km of the remnant patch (i.e. a measure of isolation). (From Webb, 1989.)

ecological interactions between the patch remnants and the surroundings, in addition to traditional criteria. This conclusion partly arose from the apparent paradoxical result that phytophagous Coleoptera point species richness **decreased** significantly with increased size of remnant patch (Figure 10.5). There was a variable plantscape at patch boundaries encouraging richness, and the small patches were particularly attractive to vagrant species coming in from the surrounding, less hospitable, agricultural matrix. The interactive edges also generated species richness, with large nearby heathland areas reducing richness (Figure 10.5). This indicates that the actual species must be assessed for their value, rather than conserving high species richness *per se*. Typical, stenotopic heathland species were found mostly in the large heathland areas, and such 'typicalness' is probably more appropriate in the evaluation process than is richness. Indeed some nature reserves are not necessarily of value for their species richness and/or diversity in comparison with the value for the rarity or typicalness of the species that they support (Figure 10.6).

Webb's (1989) findings also have implications for pest control in the agricultural matrix as well as conservation of typical species in patches. In the case of agriculturally useful, stenotopic natural enemies, large remnant (or introduced) patches are most appropriate, while for eurytopic species smaller patches may be more effective reservoirs.

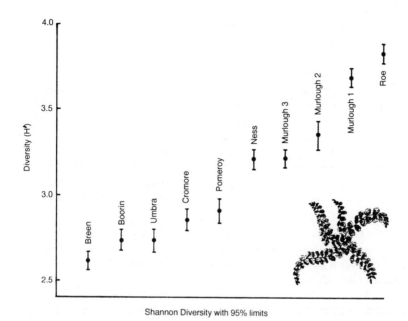

Figure 10.6 Nature reserves are not always chosen for their high species richness and/or diversity. Value of the component species may be more important, whether for rarity or typicalness. Of the 10 Irish sites measured for diversity of ground vegetation using the Shannon index H', the two nature reserves, Boorin and Breen Woods, had in fact the lowest diversity. (From Magurran, 1988.)

Once sites have been evaluated, whatever their level of disturbance, there is then the need either to preserve (without management) or to conserve (with management), remodify (in the case of agricultural and some urban land-scapes), restore and recreate sites according to the conservation goals. Management of the plantscape will generally lead to automatic management of the insect fauna. Use of burning to maintain the savanna in a plagioclimax will promote grassland insects. In contrast, setting aside of a large block of tropical forest with minimal disturbance and minimal edge will encourage a complex and diverse plantscape, while maintaining the high insect species richness with concurrent high taxonomic value.

Evaluation of landscapes does not end simply with site choice. It continues after management practices are under way, to ensure that the management methodology chosen is appropriate for the conservation goals set.

10.2 TAXONOMIC INDICATOR GROUPS

10.2.1 Indicators of biodiversity

Use of taxonomic indicator groups has two facets. A certain insect taxon may be used to identify the state or changes in a landscape. Alternatively, the goal may be to find out how certain insect taxa are affected by a possible or an inevitable modification to the landscape.

Insects in general are particularly suited for monitoring landscape change because of their abundances, species richness, ubiquitous occurrence and importance in the functioning of natural ecosystems (Rosenberg, Danks and Lehmkuhl, 1986). The class Insecta also has members, even within one taxon (e.g. Hymenoptera) that operate at different trophic levels, therefore providing varied, sensitive indication of changes.

Long-term monitoring gives variability (Samways, 1990a; Wolda, 1992), but as there is no such thing as a 'normal year' for insect abundance. A short time frame usually provides a fairly true picture without having to resort to long-term sampling (D.F. Owen and Owen, 1990). This excludes the species that occasionally outbreak to enormous population levels (Schwerdtfeger, 1941), or major range changes, and 'red shifts' in variability (Pimm and Redfearn, 1988).

Several authors have used all insect taxonomic groups in their analyses. Samples have been obtained either by a single method (e.g. fogging the forest canopy with a synthetic pyrethroid; Stork, 1991) or deliberately with a suite of capturing techniques to catch as many insect taxa (of all trophic levels), life stages, and numbers of individuals as possible to gain insight into the total insect profile of a site (e.g. Gadagkar, Chandrashekara and Nair, 1990; D.F. Owen and Owen, 1990; Clark and Samways, 1994). The problem encountered, although providing the appropriate ecological or management decision data, is that the logistics of sorting such a vast number of often unidentifiable species (especially in the tropics and south-temperate countries) makes the task formidable.

In response to this constraint, several workers have used a particular insect group, from genera through family to order. Some of the studies have rested in part on the taxonomic expertise and even preferences of the research workers, but have nevertheless provided meaningful results.

Objective choice of insect indicator groups depends on various factors. Termites would be of no value in northern Europe, but may, along with ants (Andersen, 1990), be good indicators of negative and positive factors of epigaeic environmental conditions in warm, arid areas.

Disney (1986) has suggested that Diptera plus Hymenoptera–Parasitica are useful as indicators of conservation value, as they interact with a broad spectrum of ecological niches and microhabitats. This may be more feasible in temperate areas, but in the tropics better known groups with generally large-sized species may be of greater and more expedient use in making conservation decisions. Butterflies and dragonflies are prime candidates in the tropics (Sutton and Collins, 1991). Butterflies have been targeted for temperate regions as well as the tropics (Gilbert, 1980; K.S. Brown, 1982; Pollard, 1982; Murphy and Wilcox, 1986; Erhardt and Thomas, 1991). They are generally fairly readily identifiable, there is a relatively good taxonomic knowledge of the group and they are also sensitive to environmental changes in microsite and biotope characteristics. They are often highly plant specific for their growth development (Ehrlich and Raven, 1964) and sometimes have close plant–pollinator relationships.

Pearson and Cassola (1992) have proposed the use of tiger beetles (Cicindelidae) as a good indicator group for identifying areas for biodiversity conservation. Tiger beetles are well known, their biology well understood, they occur over a broad range of biotope types and geographical areas and they also exist in remnant patches of appropriate biotopes. Interestingly, on a much smaller geographical scale, Clark and Samways (1994) also, by chance, found that cicindelids were particularly useful as 'fast indicators' of biotope quality relative to disturbance. Pearson and Cassola (1992) argue that at a site in Peru it took only 50 hours of observation to find 93% of the tiger beetle fauna, while the butterfly list of over 1200 species is still rising, and to find 90% of the butterfly species entailed nearly 1000 hours of field-work.

The importance of scale again must be emphasized. Odonata as adults may or may not be wide-ranging, and choice of scale (e.g. a site 3 m × 3 m compared with one 300 m × 300 m) influences the choice of indicator taxa. Wood and Samways (1991) found butterflies (Papilionoidea) to be good indicators of biotope type and landscape pattern at a mesoscale (e.g. 50 m × 50 m), but cicindelids were much more sensitive indicators at a microscale (e.g. 1 m × 1 m) (Clark and Samways, 1994). Further, different developmental stages give different indications, often the larva being more sensitive at the smaller scale because of its relative immobility compared with the adult. Some larvae, e.g. Odonata and Ephemeroptera, for monitoring purposes, may even be viewed as different organisms from the adults. In these orders the larvae are aquatic rather than aerial as in the adults, and represent quite different media. The corollary is that, for a species to survive, conditions must be suitable for both adult and larva. This means that presence or absence does not identify which stage in the life cycle is most sensitive, only that conditions are suitable or not for the species as a whole.

Providing the song has been related to the species, Orthoptera can also be excellent biotope indicators, as they can be recognized in the canopy at night without having to resort to any trapping or landscape disturbance (Samways, 1989f). From the viewpoints listed above, there are many variables and possibilities when choosing a taxon for environmental monitoring. It is important to define clearly the conservation question being asked, and then to choose objectively the most appropriate insect group, which may not be the one most familiar to the researcher. The contrary argument is that there is little point, unless simply morphospecies are being considered, in choosing a group that cannot be identified. If the group is also going to be used by subsequent researchers and managers, ease of recognition (e.g. butterflies and dragonflies) may be another very important consideration.

10.2.2 Indicators of disturbance

Various groups of macroinvertebrates have been used in monitoring water quality and disturbance, the Ephemeroptera and Plecoptera being particularly sensitive. Mayflies of the genus *Hexagenia* were used by Giberson, Rosenberg and Wiens (1991) for monitoring the water quality of a Canadian lake. What is of particular significance in their study is interpreting the causes of the environmental change, that is whether it is natural or whether it is man-induced, the indicator species not always making the cause clear. Certain groups show differential responses to environmental change. Erhardt and Thomas (1991) draw attention to changes in certain British Lepidoptera which were 12 times more responsive than their food plants to environmental change, while the ant *Myrmica sabuleti* was only three times as responsive.

Elliott (1991) examined the possibility of using aquatic insects as subject organisms being affected by projected climate change in Britain. Aquatic insects generally have narrow thermal requirements and vary considerably between species and between life stages of the same species. River and stream species will also be affected by possible changes in the temporal distribution of rainfall because this may influence the frequency of spates and droughts.

Elliott (1991) identified criteria that should be used in selecting freshwater species for a study of the effects of projected climate change:

1. Long-term data sets should be available on the population dynamics of the selected species and on associated climatic variables.
2. Ecophysiological information should be available on effects of climatic variables on the selected species (e.g. temperature).
3. Ecological information should be available on the functional role of the selected species within their ecosystem.

Within Britain, no species of aquatic insect could be found to meet all three criteria, but the stoneflies (Plecoptera) as a group and their eggs satisfied 2 and 3. Bearing in mind the relative extent of knowledge of the British fauna, it means that comparative work in other parts of the world, especially the tropics, requires much more basic research.

Unquestionably, insects are excellent indicators of environmental change, but the converse is that they are often readily subject to local extinction when environmental changes affect their biotope. Mobility can vary enormously even within one small taxon (Samways, 1989f). This makes fragmentation of the landscape particularly significant, reducing mobility of apparently quite vagile species, as in the case of some butterflies (Dempster, 1991). This results in certain insects restricting themselves to small patches, and with closed populations (Thomas, 1984) highly susceptible to natural and anthropogenic impacts.

10.2.3 Insects affected by disturbance

Some studies have concentrated on how landscape disturbance has affected certain insect groups. These studies make insects the subjects rather than the monitoring tools. This is an applied ecological approach with insects' welfare being the focus.

Anthropogenic landscape modification can be fairly instantaneous. Yet once the landscape is changed it can have a fairly permanent pattern of agricultural fields, plantations and urbanization. Insect behaviour is vitally important relative to these changes. Insect mobility, although not selected for coping with the appearance of a road, crop field or building, may, in the modified landscape, determine the survival or not of a species. Stenotopic low-mobility orthopteran (Samways, 1989c), coleopteran (Mader, 1984) and butterfly (Wood and Samways, 1991) species can be immediately and severely restricted by a new structure. Yet a close relative, even a congeneric, may be relatively uninfluenced, should its mobility and eurytopy, by chance, favour its survival.

The primary biological significance of dispersal is the maintenance and extension of the species in space by (re)founding and supplementing the local population (den Boer, 1990). Populations of weakly dispersing species in old landscape remnant patches are not able to spread the risk of extinction as can their eurytopic, strongly dispersing relatives in a comparatively (i.e. relative to their mobility) connected landscape. This results in populations becoming extinct in just a few years, whether they are ground beetles (den Boer, 1990), tiger beetles (Desender and Turin, 1989) or bush crickets (Samways, 1989c). This illustrates that, whatever the subject taxonomic assemblage, it is not so much differences between assemblages (with the reservation that thysanurans cannot be compared with butterflies in terms of overall mobility) but differences between species within those groups that are important. Landscape fragmentation is taking the greatest and most rapid toll on relatively immobile, stenotopic species.

10.3 INDICATOR SPECIES, SPECIES LISTS AND LIFE-HISTORY STYLES

Recording schemes of individual species in northern Europe in particular have indicated a general range reduction in species in the last few decades [e.g. Desender and Turin (1989) (West European carabids); van Swaay (1990) (Dutch butterflies)]. Indicator or flag species are worth identifying where they are associated with a particular plantscape, biotope or landscape, depending on the scale of measurement. But the paradox is that a single species which is highly sensitive to environmental perturbations may simply disappear not because of a disturbance, but because of an intrinsic feature of its population dynamics. Betting and hedging using landscape management practices may or may not save an insect species. There can be no guarantee of their continual survival, particularly for stenotopic species on the edge of their ranges (Dempster, 1991).

This brings us back to a wider base where species listing is important (Foster, 1991). Such a listing is part of Usher's (1986b) site evaluation attributes. Other attributes would also include abiotic variables. But the species list ideally should be quantitative, listing the number of species and their relative abundances (species evenness). Sites can be compared with special reference to certain key species that are characteristic of a particular biotope or landscape. Sites with several rare indicator species are of particular value, as are sites that contain several endemic species, especially if such endemics are stenotopic.

Indicator species need not always be amongst the rarest of species. Abundant species may have value, being easier to locate. However, abundant species can also be generally more eurytopic than the rarer species (Figure 7.2). This is a circular argument from one angle, because stenotopic species are bound to be rare as their particular habitat is, by definition, narrowly defined.

Certain species may also be subject to biological factors limiting their indicator usefulness. They may be highly mobile, even migratory, or they may also be subject to intraspecific and interspecific competition, and be present or absent for reasons other than landscape modification.

Species lists have definite value, and some of those species may be good indicators singly or as a group. Such shortlists must take into account the life-history styles of the species in question. Pertinent to this is the life-history style relative to the habitat templet (Southwood, 1988). Greenslade and New (1991) summarize an interesting concept in this respect. For various Tasmanian insect taxa (Table 10.2) the highest levels of endemism tended to occur in the most severe landscapes, that is those which are predictably unfavourable and in which adversity (A) selection operates (Greenslade, 1983). A-selection implies long life histories, low reproductive rates, few offspring, low dispersal ability and conservative genetic systems, all making the species susceptible to disturbance. Endemism was lowest in temporary and disturbed biotopes such as agricultural land where exploitation (r) selection for pro-

Table 10.2 Classification of some Tasmanian taxa according to level of endemism at species level, associated biotope and predominant selection type operating in that biotope. (From Greenslade and New, 1991)

Selection type	Usual biotope	Taxon	Endemism (%)
A*	Cold torrents	Psephenidae	100
	Cold, fast streams	Plecoptera	89
	Caves	Carabidae: Trechini	93
	Burrows	Parastacidae	88
	Cold, alpine lakes and streams	Anaspididae	80
	Deep soil	Oligochaeta (terrestrial)	92
	Logs (late decay)	Lucanidae	83
			91
kA†	Cold, slow streams	Tipulidae	75
	Lakes	Turbellaria (terrestrial)	69
		Trichoptera	70
		Oligochaeta (aquatic)	60
K	*Eucalyptus* forest	Coccinellidae	30
r	Pasture	Collembola	2–4

*Overall percentage endemism for A taxa = 91 (150/165 species, excluding Trechini).
† Overall percentage endemism for kA taxa = 69 (167/241 species).

ductivity was dominant and the fauna showed high population variability and high dispersal ability. Diversity was greatest in relatively stable, favourable biotopes, where interaction (K) selection for persistence in crowded, competitive environments was dominant. In conservation terms, this means that a shift from the top left corner towards the bottom in Figure 10.7 implies a loss of sensitive species, and indicates a detrimental impact upon the landscape and insect community.

Choice of a particular indicator species, as with other aspects of insect and landscape conservation evaluation depends on the precise goal, the scale of the assessment and availability of material and human resources. For determination of impact at a single localized site, a stenotopic, K-selected, endemic species (even just one of its life stages) may be appropriate. But for a large-scale survey, an abundant but biotope-restricted species may be the best tool. The goal should determine the selection of a species with an array of characteristics suitable for attaining that goal. Such characteristics may include life-history style, local or widespread abundance and distribution, availability and seasonality, sensitivity to disturbance and possibly genetic uniqueness where an historical biogeographical perspective is required.

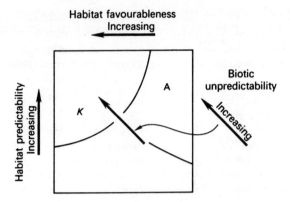

Figure 10.7 The habitat (biotope) templet with special reference to insects. A = adversity selection, r = exploitation selection for productivity, K = interaction selection for persistence in crowded, competitive, species-rich environments. (See text for details.) (From Greenslade, 1983.)

10.4 DIVERSITY MEASUREMENT

10.4.1 Practical sampling

The recurring theme in insect conservation is that size of the population and health of the species depends on conditions of the biotope and the landscape. Landscapes in turn are readily viewed from the air where features are clearly delimited. Satellite imagery has been valuable for determining landscape modification (Collins, 1990; Wells, 1989). Such images pinpoint areas of destruction, deterioration and fragmentation. These patterns can then be supplemented by ground determination of species present among the different landscape elements. This is a single time frame, and the insect species present in certain patches may not necessarily continue to survive in fragments as the years pass.

Geographic information systems (GIS) are extremely valuable for relating insect abundances to climatic and soil conditions (Johnson and Worobec, 1988; Johnson, 1989a,b) (Figure 10.8) (section 7.5).

The new technologies allow rapid mapping of point or polygon variables, the correlation of maps and the use of maps as variables in computer models. Map correlation can be used to show spatial changes in insect abundance with changes in environmental variables. When correlated with land character or landscape-use patterns, predictive computer-generated models can be produced, using programs such as SPANS (Johnson, 1989b). When certain landscapes and biotopes are identified along with significant human impacts, it is also possible to predict where the most valuable refugia might be retained from increasing surrounding degradation.

On the ground there is then the choice of sampling techniques. As there are so many insects, identification, even of a morphospecies, generally requires capture and killing of the individual. This is an ethical dilemma that has to be resolved prior to the sampling. In pragmatic terms, such sampling, unless intensively of a highly localized species, will have no detrimental impact on longer term trend of insect populations. The sampling will only cause a small ripple on the larger wave of population variation found in most insect species. Recolonization after sampling will also vary from one taxon to another (Stork, 1991) with 'reassembly' of the community gradually going through successional changes.

Sampling techniques are chosen according to the subject taxon, and no one technique will be equally effective for all taxa, even within one assemblage. One technique may not even be suitable for the different sexes of the same species or even for the same population at different times of the day or season. New (1991) discusses this problem in relation to evaluation for butterfly conservation. Fogging of the forest canopy does not dislodge female diaspid scales, yet this is a method now widely employed as a basis in estimating total world biodiversity (world species richness). More research is required on the efficiency of insecticide fogging (Stork, 1991). Similarly, other methods miss species. The false codling moth (*Cryptophlebia leucotreta*), unlike many other moths, is not attracted to ultraviolet light, while the large acridid *Acanthacris ruficornis* is not caught by sweeping, as are many other grasshoppers.

The point is that the sampling technique must be chosen after the conservation question has been posed and should be appropriate to the goals. Southwood (1978) and Gilbertson, Kent and Pyatt (1985) discuss practical sampling in general, and New (1984) critiques techniques in the context of insect conservation.

When establishing the conservation goals it is important to be clear when it is simply a spatial study or whether sampling will also continue over time. It is also important to be clear whether it is a *Q*-mode analysis (i.e. comparison and classification of sites according to their species compositions) or an *R*-mode analysis (i.e. comparison and numerical classification of species spread across the sites)(see Ludwig and Reynolds, 1988). This clarity in goal setting, and in the field and analytical methodologies, is essential. As Sutton and Collins (1991) point out, successful conservation of insects rests on adoption of simple techniques of sampling and analysis together with a small number of appropriate indicator groups.

10.4.2 Numerical diversity indices

We readily talk of biotic diversity or insect diversity when really we mean richness. Richness is the number of species present in a particular area, while

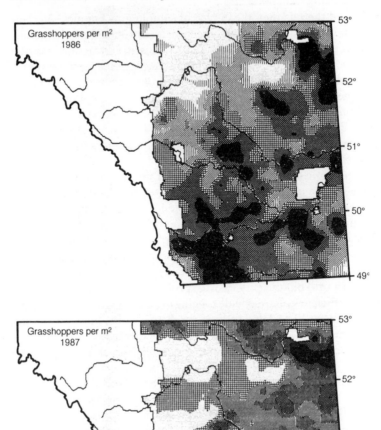

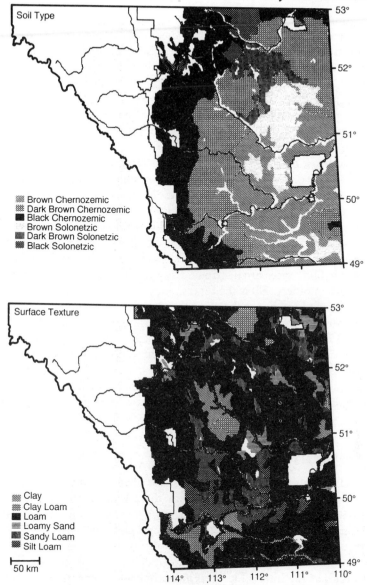

Figure 10.8 On the left are two maps of grasshopper *(Melanoplus sanguinipes, Camnula pellucida, M. packardii* and others) density in Alberta, Canada, in 1986 and 1987 based on roadside survey records. On the right are the soil classification and soil surface texture maps used in the geographic information systems overlay area analysis. An analysis of covariance, with the map intersection area as the weight, showed that, for a 10-year data set of grasshopper densities, the abundance levels were significantly related to soil type but not soil texture. (From Johnson, 1989b.)

diversity compounds richness and evenness of species (i.e. how their relative abundances are distributed). Diversity indices are useful to some extent for conservation decisions (Spellerberg, 1981; Magurran, 1988) but they have the disadvantage of being numbers with little real conceptual value. For example, the same diversity index value can be calculated for a community or assemblage with high richness and high abundance as for a comparable group of organisms with low richness and low abundance. This means that, given a particular index, the relative importance of species richness abundance often cannot be determined.

There are a vast number of diversity indices, and the interpretation of the insect community or assemblage depends on which index is used (Samways, 1984b). Different indices give different results using the same set of data. Ludwig and Reynolds (1988) propose the use of Hill's numbers (M.O. Hill, 1973), which allow greater comparability. Hill's numbers, which are in units of numbers of species, measure the **effective** number of species present in the sample. This effective number of species is a measure of the degree to which proportional abundances are distributed among the species. There are three types of Hill's numbers: $N0$ is the total number of species in the sample (i.e. richness), $N1$ measures the number of **abundant** species in the sample, while $N2$ is the number of **very abundant** species. Ludwig and Reynolds (1988) and Magurran (1988) should be consulted for calculation and interpretation attributes. A considerable degree of ecological insight can be gleaned from detailed investigations of the variety and abundances of species. A change in diversity, resulting either from a shift in the species abundance distribution or from an increase in dominance, can be highly indicative of degrees of disturbance to the landscape with components such as rivers (Magurran, 1988).

10.4.3 Graphical representation

Samways (1984b), on comparing several series of ant assemblages and using various diversity indices, concluded that the use of graphical methods, especially rank–abundance curves, is far more meaningful in interpretative terms. Plotting the relative abundances on a graph gives an instantly clear, visible picture. Such graphical representation can always be supplemented with one or more appropriate multivariate analysis techniques operating on named species at the interval level of measurement. However, assemblages and communities are not static: the temporal component, whether seasonal or longer-term, changes the spatial interpretation. This is especially so with insects, which often have high natalities, mortalities and mobilities.

10.4.4 Multivariate analysis

A further point of importance in the use of classification using multivariate

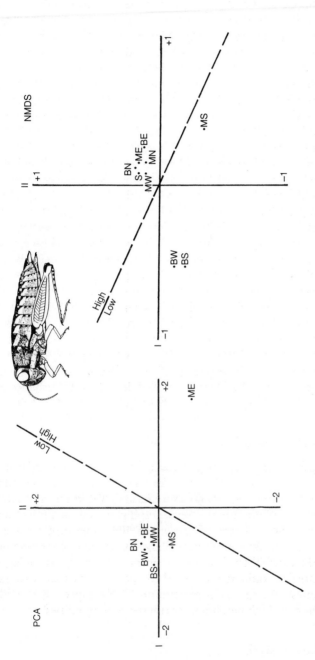

Figure 10.9 Results of using two different ordination methods, principal components analysis (PCA) and non-metric multiple dimensional scaling, (NMDS) on a grasshopper assemblage associated with hilltops in a reserve area in South Africa. PCA clustered the six sites with lowest grasshopper density, while NMDS clustered the six sites with highest densities. The techniques did not contradict each other, but gave a different complexion. S = summit, MN = mid-slope (north), BN = base (north), ME = mid-slope (east), BE = base (east), MS = mid-slope (south), BS = base (south), MW = mid-slope (west), BW = base (west). (From Samways, 1990b. Reprinted by permission of the Society for Conservation Biology and Blackwell Scientific Publications.)

techniques is that it may be illuminating to use more than one technique to give a slightly different complexion to the conservation question. For example, in the ordination of biotopes of montane grasshopper assemblages in southern Africa, principal component analysis tightly clustered sites with the lowest density of grasshoppers, while non-metric multiple dimensional scaling clustered the sites with the highest density from the other sites (Figure 10.9).

Many techniques are available (see Gauch, 1982; Pielou, 1984; Manly, 1986; Digby and Kempton, 1987; Jongman, ter Braak and van Tongeren, 1987), but two that have gathered favour in recent years are detrended correspondence analysis (DECORANA) and two-way species indicator analysis (TWINSPAN). DECORANA, by ordinating sites and species, allows the major ecological factors that determine variation to be identified, while TWINSPAN produces a classification of site lists based on an ordered two-way table of sites and species with each division into subgroups characterized by indicator species. This technique has been successfully used for water beetle site list evaluation (e.g. Foster, 1991). Jongman, ter Braak and van Tongeren (1987) also recommend canonical community ordination (CANOCO), a form of canonical correspondence analysis, which performs a DECORANA ordination, after which it determines the environmental variables that determine the axes. This method has considerable merit as it can perform the ordination, and, using the same program, relate the results to environmental variables.

Whether species richness, species diversity, rank–abundance curves or ordination are used, all these methods are only descriptive. They do not explain the reason for the species patterns. For explanations, descriptions of species in the landscape need to be followed up by experimental verification using scientific methodology, as was done for grasshopper conservation (Samways, 1990a).

10.4.5 Beyond numerical diversity measures

An inherent problem with diversity indices, rank–abundance curves and multivariate analyses in conservation biology is that they are quantitative, with all species, taxonomically and functionally, being implicitly treated as equal. With the crisis science of conservation biology requiring urgent choices to be made, species are not equal, whatever valuing system is used. Qualitative decisions have to be made alongside quantitative ones.

Some quantitative decisions automatically encompass quality. Preservation of rain forest tracts no doubt protects many rare endemic insect species. Protection of the same area of forest in Britain would not protect the same amount of genetic uniqueness, although geographically different human values would come into play, with forest protection leading to insect species preservation of a different kind.

Going beyond numerical diversity indices and beyond regional human value systems, biodiversity and insect conservation must recognize the phylogenetic value of species. Additionally, it must recognize their functional attributes within ecosystems (Cousins, 1991). The functional aspect relates to how species contribute to community structure and ecosystem function, and to the relative sensitivity of species to landscape change. These points are discussed in the next two sections.

10.4.6 Phylogenetic diversity measures

Determining taxonomically valuable species requires a knowledge of evolutionary lines. This involves identifying the phylogenetic uniqueness of a species or group of species (Wheeler, 1990). This phylogenetic knowledge must be acquired as fast as possible, as the rate of species and habitat destruction is increasing. This knowledge, gained quickly, will be part of the conservation evaluation and management process. Gained slowly, it will then only be on pinned specimens *in memoriam* of their now extinct forebears who lost their habitats.

Vane-Wright, Humphries and Williams (1991) point out that critical faunas analysis (sections 3.4 and 10.1) does not offer a sufficiently flexible strategy at the global level. The critical faunas approach prioritizes sites (countries), selecting one, then choosing the next more valuable one from the remaining sites. The procedure is based on endemics. A better strategy for obtaining higher average diversity scores per site is possible if multiple-site choices are considered. This involves selection in one step of a priority set of sites which together make up maximum diversity quantitatively and qualitatively.

By applying a qualitative taxic weighting, a phylogenetic diversity measure can be obtained where sites complement each other in species without overemphasis on endemics. This approach could be applied at scales of measurement from local patches or landscapes to whole regions or countries.

Vane-Wright, Humphries and Williams (1991) illustrate their method with reference to 43 species of bumblebees in the *Bombus sibiricus* group. They used a hierarchical taxonomic classification to determine areas around the world of taxonomic distinctiveness (Figure 10.10). The classification is based on an information measure determined by the number of branch points in the classification dendrogram [see Vane-Wright, Humphries and Williams (1991) for details of calculation]. The reasoning behind the approach is that if there are two threatened taxa, with one of them not closely related to another, widespread and common species, it is reasonable that highest priority is given to the taxonomically most distinct or remote form (Atkinson, 1989). The *B. sibiricus* group analyses illustrate that, if simple species counts are used to locate the unit area of maximum diversity, then Ecuador with its

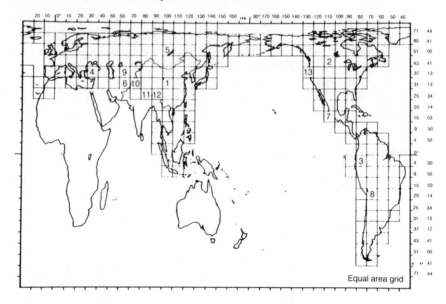

Figure 10.10 Identification of areas by the taxonomic distinctiveness of their bumblebee fauna of the *Bombus sibiricus*-group. Highest successive scores are given to taxically weighted (i.e. distinctive) complementary diversity. The areas are labelled by reference to major geopolitical features near the centre of the particular grid squares, as follows: 1, Gansu (China); 2, Big Horn (USA); 3, Ecuador; 4, Turkey; 5, Baikal (Russia); 6, Hindu Kush (Afghanistan); 7, Michoacan (Mexico); 8, La Paz (Bolivia); 9, Samarkand (Russia); 10, Kashmir (India); 11, Nepal; 12, Arunachal Pradesh (India); 13, northern California (USA). (From Vane-Wright, Humphries and Williams, 1991.)

10 species (15% of the total weighted diversity score) is of highest priority. But when taxonomic distinctiveness is calculated, Gansu in China is of more immediate concern, with its nine higher ranking species, contributing almost 23% of the total weighted diversity score. When one area is used to complement another (i.e. the percentages added), the 13 highest percentage areas make up 100%. Conservation of these 13 'first' areas would theoretically conserve all 43 bumblebee species (Figure 10.11).

May (1990a) suggests a refinement of the method of Vane-Wright, Humphries and Williams (1991), in which hierarchical trees are only partially resolved, and where some branches are binary, while others are multifoliate. The suggestion is to count not simply the number of nodes between tree terminals and root, but rather sum the branches at all such nodes.

Nixon and Wheeler (1992) give two simple measures of phylogenetic diversity that are easy to apply to circumstances where decisions must be

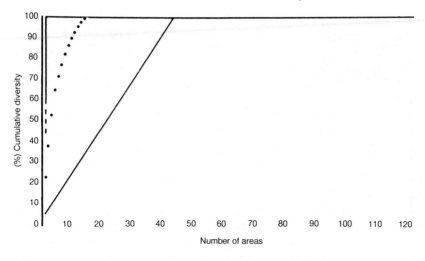

Figure 10.11 Weighted taxic diversity accumulation histogram for the 43 bumble-bee species of the *Bombus sibiricus* group. The abscissa represents the number of sample areas (the *sibiricus* group occurs in 120 of the 250 grid squares shown in Figure 10.10). The ordinate represents percentage diversity (based on the aggregate classification). The 13 points correspond to the sequence of grid squares listed in the legend to Figure10.10, the 13 priority areas. Thus the first point is area 1, Gansu, contributing 22.95% of the total weighted diversity; area 2 is Big Horn, contributing an additional 15.25% of complementary weighted diversity, and so on up to area 13, northern California, which contributes the last 2.19% of weighted diversity, and brings the accumulation to 100%. The (incomplete) vertical straight line in the top, left corner indicates how the histogram would look if there was at least one grid square containing all 43 species. The sloping straight line is a bound indication of how the graph would look if all 43 species occurred in different grid squares and all were treated as taxonomically equal. (From Vane-Wright, Humphries and Williams, 1991.)

made to collect, study or conserve one or another taxon. The two indices, the unweighted binary phylogenetic diversity index (BPD) and the weighted phylogenetic diversity index (WPD), rank species according to the amount of phylogenetic diversity in the clades to which they belong. The application of these measures does not require complete resolution of species relationships in the hierarchical tree. This is particularly valuable in the conservation context, where fully determined phylogenetic trees are rarely likely to be available. The method is based on the number of species subtended by each node in the hierarchical tree. A matrix is composed in which the rows are converted to decimals to the base 2. The decimal values are then ranked [see Nixon and Wheeler (1992) for details of calculation]. The outcome of the BPD is that the ranking accords all members of a branch the same priority relative to all members of its sister branch. Thus, in a classification based on phylogeny, when comparing two genera, all species of one genus will have a higher rank than all species of the other genus, although the species within each genus may have different ranks relative to each other. While the BPD is probably the best estimate of relative phylogenetic uniqueness, it does not disperse the rankings so that the **overall** sample of phylogenetic diversity is maximized. As Nixon and Wheeler (1992) put it, using the BPD all orchid species may be ranked lower than all species of liverwort.

There can exist no measure which maximizes both uniqueness and breadth of sample, although the WPD is a compromise towards these two goals. Nixon and Wheeler (1992) describe the calculation of the WPD; the measure ranks the suite of species, rather than giving an absolute measure of phylogenetic diversity. The WPD provides a way of estimating the status of higher level phylogenetic groups. Priorities based on such estimates are useful, as the loss of the last member of a branch should be considered a greater phylogenetic loss than the loss of a single species of its sister branch, if that branch is much more diverse. In short, the WPD provides a balanced ranking for priority decisions, especially concerning individual species.

In contrast, the BPD may be more appropriate when considering geographic or area priorities. It might be used by ranking all species at the sites being considered, then scoring each site for a composite or average rank. This would provide an estimate of the relative phylogenetic uniqueness of the sites.

As with numerical diversity measures, phylogenetic diversity measures require a good database. The basic information may not be available, as in the case of the majority of tropical forest species. The selection and interpretation of the phylogenetic diversity measures must also be in accordance with the precise conservation goals. Over the next few years, there is likely to be a proliferation of such measures, from which the practising conservationist will be able to choose the most useful measure for the purpose at hand (Williams, Humphries and Vane-Wright, 1991). At the present time, the emphasis is mostly on phylogenetic uniqueness. This may not be the only approach, with the concepts of 'representativeness', 'typicalness' and 'rarity' applying as much to phylogenetic conservation evaluation as to landscape and ecosystem conservation evaluation.

10.4.7 Sensitivity to landscape change and species functional roles

With landscape modification, habitat destruction and global climate change, the biotic loss mostly will be through the extirpation of restricted-range, stenotopic species. Those that can survive in changed and changing ecosystems will be the ones most likely to survive in the future.

The problem with diversity measures of any type is that they do not tell us anything about the responses of insects, other animals or plants to changing conditions. The circular problem is that we need to know all the responses of all the species under all conditions, but there is not time for this to save them all.

Bearing in mind the need to conserve both for rarity (especially range-restricted endemics) and for typicalness, sites have to be selected that characterize all types of ecosystems and landscapes. Other factors interact, particularly the significance of shape, size, remoteness, fragmentation and adversity risk of sites. Most of the ecosystem patterns prior to man's impact were driven principally by climate, both historical and contemporary. The climatic patterns today are therefore of great value, particularly as seen by the vegetation types of the earth (Walter, 1985). Preservation of the vegetation patterns **and** processes, static structure and dynamic succession will provide support for most insects. This is not so much a shotgun approach as a broadbrush, given the short period of time in which to act.

The crux of the present global problem is not only what there is and where it is on the earth's surface, but also what is going to happen to it, in view of the surface and air small-scale and large-scale changes. Although there is an implicit assumption that many species will become extinct, we have no real foundation for predicting what this loss will actually be, or what will be the consequences.

An important underpinning concept is to recognize the **function** of the organisms. Insects, like the micro-organisms, are an understudied yet vitally important functional component of ecosystems. Chapin, Schulze and Mooney (1992) have drawn attention to functional equivalence between species. There are two extremes of functional equivalence:

1. Each species in an ecosystem plays a relatively vital role, and removal of each species weakens the integrity of the ecosystems, just as removal of rivets from an aeroplane weakens its structure (the rivet hypothesis).
2. An ecosystem is composed of partially overlapping functional groups, each with ecologically equivalent species, so that some species can be lost from the communities with little effect upon the ecosystem processes (the redundancy hypothesis).

As Chapin, Schulze and Mooney (1992) point out, the truth probably lies somewhere in between. Loss of a particular plant species may cause extinction of certain insect species but the ecosystem continues more or less intact. Yet loss of a particular insect species may have no detectable impact on ecosystem function.

Redundancy can be a species-packing phenomenon: lack of adverse conditions may allow a multitude of species to evolve, where only one may have evolved under more stressful environmental conditions. Any environmental change will knock out species, one by one, the most sensitive species going first. If there is high redundancy among species in a functional group, the community will, at first, continue unchanged. This will be the case at least from a functional point of view, if not structural.

A keystone functional species is a unique pivotal organism without redundancy. Addition or removal of a keystone species may cause major changes in community structure and ecosystem processes. One of the challenges of biodiversity conservation is to predict which are these keystone species, although it may be difficult to predict the exact repercussions of their addition or removal. Also, the maintenance of ecosystem integrity depends on the environmental sensitivity of functional keystone species. A sensitive, stenotopic keystone species will need particular conservation attention.

Evaluation methodology needs now to research the significance of biotic diversity and what the critical thresholds for change are when certain species of plants, insects or other biota are added or subtracted from the system. The global climate changes now taking place, and the value of insects as experimental animals, provide us with an observational window to observe how different populations, species, ecosystems and landscapes (anthropogenically modified and unmodified) will respond to the inevitable changes.

10.4.8 Temporal variability

Identification and description of sites implies some temporal freezing. Insects are often highly variable in their population levels, and communities are dynamic. Survival of insect populations implies both spatial and temporal factors. Forney and Gilpin (1989) studied *Drosophila* spp. mobilities and survival in different size and shape biotope patches. They found that a single large biotope patch showed a lower extinction rate than the combined rate in two small patches of the species. The habitat corridor between the pair of small patches seemed to produce a 'rescue effect' that lowered extinction rates. This was probably the result of a decrease in the coefficient of variation in fluctuations of the population size in the coupled system.

Some autecological studies on butterflies have shown that even highly localized and threatened species undergo major population variations (Ehrlich and Murphy, 1987) (Figure 10.12). During such fluctuations there may be changes in behaviour. The bay checkerspot butterfly *Euphydryas editha*, for example, forms mating aggregations at low densities which probably play a crucial role in keeping a population viable (Ehrlich and Murphy, 1987). However, at high densities males move to hilltops. The important point is that certain features of the environment may seem unimportant in one year of observation, yet in another year, when densities are different, other features may play a role.

Environmental stochasticity can also be highly significant in exterminating small populations, emphasizing the importance of preserving several populations. The long-term studies that have been carried out on butterflies in

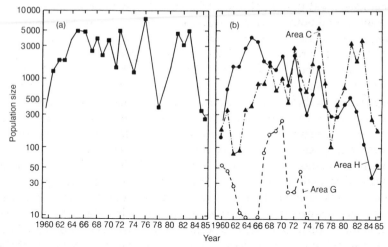

Figure 10.12 Population fluctuations in a small colony of the threatened bay checkerspot butterfly (*Euphydryas editha*) in California. (a) The overall population fluctuates considerably from one year to the next. (b) When the population is divided into separate demographic units, it can be seen that in area G, it became extinct, then recolonized and became extinct again. This illustrates, when areas C, G, and H are compared, that there is considerable difference in demographic trends, even in nearby areas. (From Ehrlich and Murphy, 1987. Reprinted by permission of the Society for Conservation Biology and Blackwell Scientific Publications.)

both the USA and the UK have emphasized that reserves may need to be bigger than normally perceived. But even large populations do not guarantee survival of the subspecies or species, as seen by the orthopterans *Decticus verrucivorus monspeliensis* in France (Samways, 1989a) and *Melanoplus spretus* in the USA (Lockwood and De Brey, 1990), which are now extinct yet were once so common as to be pests.

Ehrlich and Murphy, (1987) point out that a minimum viable metapopulation approach may be necessary for conservation directives. This is particularly so in view of how readily isolated populations of insects can become extinct (Erhardt and Thomas, 1991). These are important pointers for the tropics, where time is too short for long-term single-species studies. To summarize, temporal studies clearly indicate that as large as possible tracts of land must be conserved. An apparent irony however, is that despite the fact that about 95% of the Atlantic lowland Brazilian forest has already gone, no butterfly extinctions have been recorded (K.S. Brown, 1991). This may be because the species are so rare that it would be difficult to know whether they have disappeared or not. But the warning from studies in temperate lands is that although there is cause for some optimism, with man living alongside many of the species (K.S. Brown, 1991) in the shorter term, longer term projections are less optimistic, as population by population becomes extinct.

10.4.9 Temporal variability measurement

Variability and its measurement are particularly important as they may well be related to the fact that it is not unusual for local insect species to go extinct (Dempster, 1989). This dynamic characteristic overlays the spatial distribution of species, and may relate to recent concepts on the geographical distribution of rarity (Schoener, 1987, 1990)(section 7.1).

As insect population data do not usually follow a statistically normal distribution, it is usually necessary to transform the data (see Southwood, 1978). This applies often to spatial as well as temporal data. Standard deviation $\log_{10}$ population size has been widely used in community **stability** studies (Connell and Sousa, 1983; Schoener, 1986; Pimm and Redfearn, 1988; Owen and Gilbert, 1989). But for **comparative variability**, it is not generally independent of the mean, especially for widely differing population densities.

Very little work has been done on the quantitative variability of insect populations [but see for example Wolda (1988, 1992) and Owen and Gilbert (1989)], although Williamson (1972,1984) has discussed general principles.

A detailed comparison of five measures of **variability** showed that deter-

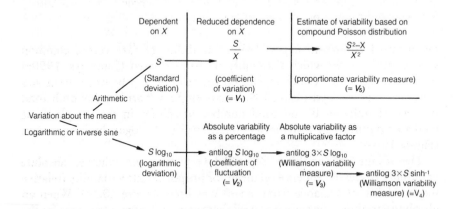

Figure 10.13 Some possible temporal population variability measures and their relationship with each other (see text for discussion). (From Samways, 1990a. Reprinted with the permission of Springer-Verlag, Heidelberg.)

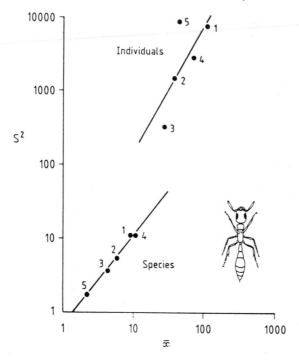

Figure 10.14 Regression of temporal variance of southern African ants on mean for individuals and for species in point sites in five different biotopes. The species numbers vary because some become extinct and reappear in the experimental quadrats. Individuals: $S^2 = 3.036x^{1.691}$; $S^2 = 0.677x^{1.224}$. (From Samways,1990a. Reprinted with permission of Springer-Verlag, Heidelberg.)

mination of comparative population variability of rare versus common species depends on which variability measure is used (Samways, 1990a) (Figure 10.13). The results depend on the relationship between mean and variance. Firstly, it was established that temporal **variance** for both total number of individuals and total number of species in the sample areas increased significantly with mean population size by a logarithmic regression (Figure 10.14).

The results on **variability** depended very much on whether **absolute** variability (some dependence on mean) or **proportionate** variability (relative independence of variance from mean) was used (Figure 10.12). When an absolute variability measure was used there was an apparent tendency for the populations of common species to be more variable than either mid-range or rare species (Figure 10.15). This was not the case when the proportionate variability measure (V_s) was used, with rare species being just as proportionately variable as common ones (Table 10.3).

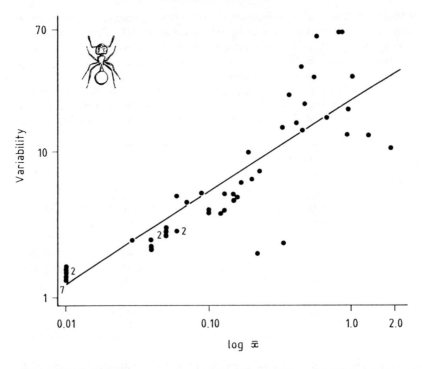

Figure 10.15. Absolute temporal variability (V_3; see Figure 10.13) in relation to mean number of southern African ants present for each species in five different biotopes. Each plotted point represents a species. The regression line corresponds to the power relation $y = 22.10x^{0.632}$. Note that there is greatest scatter of variability among the most common species. For further discussion see text, and see Table 10.3 for a comparison with the proportionate variability measure (V_5). (From Samways, 1990a. Reprinted with the permission of Springer-Verlag, Heidelberg.)

When different variability measures were compared, they not only behaved differently from each other, but also varied depending on whether only the common species (i.e. first quartile) or increasingly rare species were included in the analysis (i.e. up to quartiles 1–4) (Table 10.3). This was particularly so for measures V_3 and V_4, which also performed poorly at low population densities, suggesting that the results were more reliable when only the common species were included. When the proportionate variability measure (V_5) was calculated, no matter which quartiles were used, no one group (common, rare, etc.) of species was more, or less, variable than any other group. This suggested that extinction could happen to any population (or species) no matter what its average population level. This seems to have been borne out not only by rarities such as the large blue butterfly (*Maculinea*

Table 10.3 Linear correlation coefficients, and levels of significance, of five variability measures (see Figure 10.13). V_1 = coefficient of variation, S/X; V_2 = coefficient of fluctuation, antilog $S \log_{10}$; V_3 = Williamson variability measure, antilog $3 \times S \log_{10}$; V_4 = Williamson variability measure on inverse sine transformed raw values, antilog $3 \times S \sinh^{-1}$; V_5 = proportionate variability measure, $(S^2 - X/X^2)$, on mean population density for (1) the 14 most common ant species (first quartile), (2) the 28 most common species (first and second quartiles), (3) the 42 most common species (first, second and third quartiles) and (4) all 56 species (all four quartiles) of southern African ants. (From Samways, 1990a)

Variability measure	Quartiles 1	1–2	1–3	1–4
V_1	$r = -0.56$	$r = -0.43$	$r = -0.38$	$r = -0.25$
	$P<0.05$	$P = <0.05$	$P = <0.02$	$P<0.1$
V_2	$r = -0.31$	$r = -0.08$	$r = -0.20$	$r = -0.26$
	$P>0.1$	$P>0.1$	$P>0.1$	$P>0.1$
V_3	$r = -0.26$	$r = 0.33$	$r = 0.48$	$r = 0.54$
	$P>0.1$	$P<0.1$	$P<0.01$	$P<0.001$
V_4	$r = 0.020$	$r = 0.21$	$r = 0.30$	$r = 0.34$
	$P>0.1$	$P>0.1$	$P<0.1$	$P<0.01$
V_5	$r = -0.26$	$r = -0.17$	$r = -0.18$	$r = -0.12$
	$P>0.1$	$P>0.1$	$P>0.1$	$P>0.1$

arion) in Britain but also by the now-extinct formerly common orthopteroids, *Melanoplus spretus* and *Decticus verrucivorus monspeliensis*.

It is well known that certain species of insects can remain at low population levels for many years and then suddenly escalate (Schwerdtfeger, 1941). But generally it seems that most rare species of insect remain at steadily low mean population levels.

A first premise is that rare species present analytical problems when quadrat size is predetermined. A set quadrat size inevitably means that there may be zero counts, precluding any numerical analytical procedure. Besides analysis and evaluation, rare species also present conservation management difficulties. Should harsh environmental or even biotic conditions (e.g. suddenly more competition) become increasingly intense, the rare species are likely to suffer more than common ones in the face of absolute density-independent mortality. This is further aggravated when their biotope is fragmented. In short, small, isolated populations of insects have little chance of survival in the long run, as illustrated by Ehrlich and Murphy (1987), Dempster (1991), Samways (1989f) and J.A. Thomas (1991).

10.5 SUMMARY

Whether from an ethical or pragmatic viewpoint, insect individuals, populations, species, and all the interactions, are generally protected when the landscape is conserved. The relatively large scale of the landscape with its fairly well-definable structural composition makes it a good umbrella for preservation of smaller scale units and processes.

The landscape interacts with larger scale historical and contemporary processes, such as past and present climatic events, to give the landscape its pattern and functional character. Species that occur across a whole region can become extinct because of widespread events coupled with fragmentation of the landscape. Scale of measurement, temporal or spatial, has an important bearing on interpretation of temporal events and spatial patterns. Landscape conservation protects the insects, along with other biota and geology. In turn, insects themselves may be used in evaluation methodology for conservation of landscapes.

On the global scale, the identification of hotspots, wilderness areas, megadiversity countries and critical faunas has been used to identify areas particularly rich in insect species, especially endemics. At the landscape and lower levels, it is important to apply a rigorous scientific approach of hypothesis testing. This is associated with the setting of specific conservation management goals. As conservation biology is a many-faceted field and time is short, the introduction of workable evaluation methods is essential.

Measurements of richness and diversity are principally quantitative, with little value being attached to the differences between species. Yet species differ taxonomically and functionally, as well as in different levels of rarity. Sites may be selected on the basis that they contain many rare species, or because they are home to localized and rare endemics. Sites may also be selected because they are species-dynamo areas, which, over evolutionary time, will contribute to future diversity. Site-selection criteria include 'naturalness' (freedom from disturbance by man), 'typicalness' (a typical example from a region characterized by certain ecosystems and landscapes) and 'representativeness' (a series of sites are selected and ranked according to how they represent the ecosystems of the area).

After evaluation and reserve selection, there is the need to manage the sites with an awareness of the ecological interactions taking place. The edges between patches and the matrix determine not only numbers of species, but also types of species, taxonomically and behaviourally. These ecotonal influences also have a bearing on conservation in agricultural landscapes and in the optimal use of natural predators for controlling pests. Even after site selection, it is necessary to re-evaluate periodically to determine whether management practices are appropriate to the goals set.

Various insect taxonomic groups are useful in identification of particular landscapes or communities for general conservation of biota, or for conservation of the group itself. Various taxonomic groups, from butterflies or dragonflies to tiger beetles or flies, have proved useful in various contexts.

Some studies have used rather the total insect fauna caught with a suite of capturing techniques. Other studies have used a specific sampling technique, such as fogging with an insecticide, to sample a wide selection of the susceptible insect fauna. Certain insect groups, such as British stoneflies, appear suitable for monitoring global climate changes.

Insects themselves are affected by disturbance to the landscape. With the large variety of species with various responses, and the large number of individuals, insects have proved to be good environmental monitors. They have also been good subjects for examining how landscape pattern change affects faunas. Some individual species are good indicators of a particular facet of change in the landscape, although evaluations with single species do only provide a very narrow assessment window.

Different sampling techniques have different limitations, as do the many analytical techniques for measuring diversity. Although numerical diversity measures, multivariate techniques and graphical methods are useful for quantifying diversity, they do not provide a qualitative assessment. Species are not equal in their value. Phylogenetic diversity indices are now being developed to weight the value of species according to their taxonomic distinctiveness. The functional value of species is also important in biodiversity evaluation, with much more research now being required to determine the sensitivity of species to landscape change. There is also the need to establish the functional equivalence of species, to know which species may be lost from the system with little impact on community structure and ecosystem processes (i.e. species with high redundancy) and to know which species are keystone species whose loss would have major impacts (i.e. species with low redundancy).

Temporal variability is particularly important in insect conservation biology, as insect populations fluctuate so much. But temporal variability measurement is fraught with analytical difficulties depending on whether **absolute** variability (variance influenced by the mean) or **proportionate** variability (variance relatively independent of mean) is used. The suggestion from the results of variability measures is that any species can become extinct no matter how abundant it is, should conditions go against its continued survival. Indeed, the Rocky Mountain grasshopper, which in the middle of the last century was so common and a serious pest in western USA, was extinct by 1902.

-11

Stopping the loss of individuals, populations, species and landscapes

Calculate threats, assess risks, develop a strategy to provide the optimum combination of opportunities.

Erwin Rommel

... radical ecotherapies may peak in the temperate zone around the year 2100. We cannot be so sanguine about the tropics, where the demographic winter will be more severe and longer.

Michael E. Soulé

11.1 RESTORATION ECOLOGY

11.1.1 Positive, active optimism first

Michael E. Soulé, one of the greatest contemporary conservation biology thinkers, suggests that, as technological change is occurring at such an accelerating rate, we might as well go surfing, as try and do something for conservation of wildlife into the twenty-first century (Soulé, 1989). This is highly symbolic. Now it is timely for the new generation to take over, whether Earth First(!)ers or our children and grandchildren. This generation of conservation scientists have included doomsday alarmists risking scientific credibility for warning of imminent changes. The world needs both sides: hard tack empirical scientists and futurists. Both are of value for conserving species and landscapes. But we cannot afford pessimism. Our cerebral cortex offers us the choice of bioempathy or not. We go forward with banners of both reductionism and holism, with a concern for nature and for mankind.

There is no evidence, strictly scientifically speaking, that collapse of ecosystems, if such a thing exists, will lead to human detriment. The earth will, though, lose species. Keystone species aside, this is an emotional issue rather than a rational one. The problem is that we crave manicurism, from housing complexes to bowling greens. Insects are often seen as the antithesis of this. Yet they represent the creative variety of life that makes the earth so different from Mars. This is not that far away planet circling Castor. This is not entomological fanaticism, but simply restating Briggs' (1991) observation that much of the earth's diversity is terrestrial. Land, making up only 29% of the earth's surface, embraced by water and air begets the richest tapestry of biodiversity. Now given our conscience, what do we do about conserving it?

11.1.2 A first premise

We cannot just go surfing–not with our moral fibre and resourcefulness. Maintaining as many biotopes and as many landscapes as possible, underpinned by an understanding of population genetics and demographic processes, is essential, ethically and pragmatically. Manager researchers, particularly landscape ecologists, must provide the guidelines. The principle is simple. Conserve as many large and heterogeneous patches (managed where necessary) as possible (Moore, 1991b). Ideally, these patches should be remnant or resource, and be connected. They may even be introduced where suitably chosen. This leads to the healing of damaged landscapes through restoration activities.

11.1.3 Recreating and restoring landscapes

Where the plant cover has been completely destroyed, and even the topography modified, as with mining activities, the biotopes may have to be completely recreated. Natural ecosystems damaged from other major disturbances such as deforestation and agricultural modification need to be restored. This field of restoration ecology is multidisciplinary (Jordan, Gilpin and Aber, 1987; Buckley, 1989). To date, most restoration has taken place in the north-temperate lands, where some projects have been devised with insect conservation as an integral part. These projects are becoming increasingly large, with the largest man-made nature reserve of 2500 acres now being established at Teeside, northern England.

Restoring sites for insect conservation is intimately associated with the establishment of the plantscape in which the insects live. Introduction of certain plants is automatically followed by colonization by many other plant species. At a landfill site in England, Davis (1989) established 27 plant species, but another 140 species colonized naturally. However, four of his total of 31 sown plant species failed to establish.

Of particular interest are the ecological restoration activities in the tropics, where endemism is high (Janzen, 1989). The transformation of the Amazonian moist tropical forests has produced a mosaic of pastures and regrowth forests which are agriculturally unproductive, biologically impoverished, and far more flammable than the mature forests that they replace. During recovery, biomass and species accumulate at a rate that is inversely related to the intensity of use prior to abandonment, with slowest regeneration on bulldozer-scraped land. The grass and shrub-dominated old fields resist forest regrowth because of various inhibitory factors on tree establishment and growth, such as low propagule availability, seed predation, seedling deprivation, seasonal drought and root competition with old field vegetation. The need now is to establish tree-based agricultural systems and to restore the forest regenerative capacity in old fields by planting trees that attract seed-carrying animals and ameliorate harsh environmental conditions (Nepstad, Uhl and Serrão, 1991).

In Britain, the recreation of floristically rich grassland, which encourages a variety of insect life, can be achieved using several methods (Wells, 1989). When establishing grasslands from seed, it is important to choose the right seed mixture, to use a nurse cover (e.g. Westerwolds rye grass) and to carefully prepare the site and seed bed. Seeds selected should meet the following criteria (Wells, 1989): they should be ecologically suitable for particular soil/water conditions; they should be common grassland species, not rare or locally distributed species, and preferably be perennial and long-lived; they should have colourful and attractive flowers, be attractive to insects as nectar or pollen sources and they should not be highly competitive or invasive; their seeds should germinate over a range of temperatures.

At a restored site, insect succession follows the plant succession as the biotope changes. Also, taxonomic and trophic insect diversity has been found to increase with plant succession (Southwood, Brown and Reader, 1979; Brown and Southwood, 1983). In a 'young field site' (1–2 years abandoned) it was found that 93.0% of insect species were herbivores, 5.5% predators and 1.5% fungivores, while in an 'old woodland site' (about 60 years since abandonment), 74.6% were herbivores, 10.0% were predators, 14.6% fungivores and 0.8% scavengers.

Davis' (1989) study was characterized by widespread species of butterfly readily colonizing his site. These were all highly mobile species and arrived before the plots were sown. There was therefore no barrier to their entering the recreated biotopes. This is an important feature, because even apparently strong insect fliers can be readily inhibited from moving over unsuitable habitat, even only 100 m wide in the case of the adonis blue butterfly

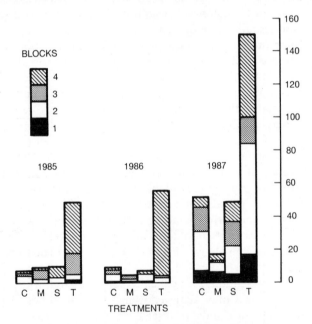

Figure 11.1 Population levels of the common blue butterfly (*Polyommatus icarus*) on using four ecologically restoring treatments (C = control plot which was disturbed periodically by cultivation, but not sown; M = a treatment area with an elongated mound of fine limestone; S = short-growing sown mixture of grasses and wildflowers; T = tall-growing sown mixture of grasses and wildflowers). There were four replicated blocks making up each treatment. In all three years, and combining blocks, the species distinctly preferred the tall mixture treatment. (From Davis, 1989.)

(*Lysandra bellargus*) (Thomas, 1983a), and far less than that in some southern-hemisphere species (Wood and Samways, 1991).

11.1.4 Assembly rules

In Davis' (1989) study, the dispersal inhibition in butterflies was highly significant and strongly affected the assemblage pattern. After appropriate site management, such as sowing of grass, fertilizer applications and creation of patches of bare ground, two species, the marbled white butterfly (*Melanargia galathea*) and the grayling butterfly (*Hipparchia semele*), were introduced. Long-term monitoring is now required to determine the success of these introductions.

Davis' (1989) experiments illustrate the value of detailed autecological studies when specific species are targeted for conservation. This is seen in Figure 11.1, which shows that the common blue butterfly (*Polyommatus icarus*) preferred the tall-grass plots which also contained its food plant, bird's-foot trefoil (*Lotus corniculatus*). Without such a characteristic plantscape, the assembly pattern would have been different.

Gilpin (1987), working on experimental laboratory cultures of *Drosophila*

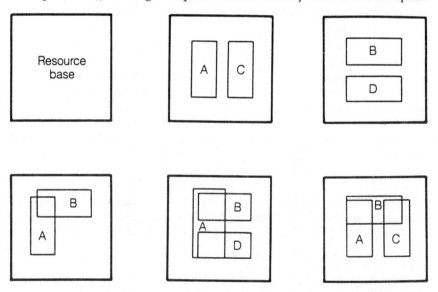

Figure 11.2 A conceptual representation of several species depending on the same resource base. They may coexist if they exploit the resource in different ways and so do not compete. Combinations are ruled out, but the two pairs, A and C and B and D, which can coexist, represent alternative stable combinations. A knowledge of these types of interactions determine the success or otherwise of restoring a community. (From Gilpin, 1987.)

on a tight resource base, verified that there do seem to be assembly rules for communities in the formation (Figure 11.2).

Competition, order of introduction and identification of keystone species were important factors determining the outcome of these synthetic restoration experiments. Under field conditions such factors may be less important as the biotope is so much more complex structurally and functionally and subject to the vagaries of weather. Competition may subside as plant architecture increases, and as pathogens, parasitoids, predators and mutualists, as well as resource availability, change both spatially and temporally. Practical restoration conservation may turn out to be more of an art than a science, yet the subject provides great opportunities for scientific study.

11.1.5 Translocation, reintroduction and re-establishment

Movement of individuals from one locality to another is **translocation**. The receiving locality may or may not have supported the species in the past. Some translocations may involve boosting the population level of the species in the receiving locality. **Reintroduction** uses individuals of a species from either captive-bred stock or another wild locality (i.e. translocation), to re-establish a species that was formerly present but locally has died out. These methods have moral and ecological implications.

Any translocation or reintroduction is an anthropogenic impact, albeit to correct an adverse situation often created by man. Ecologically, the outcome of a translocation or reintroduction is difficult to predict. Where the species is not a biocontrol agent or a keystone species, there is likely to be high ecological redundancy, and the exercise will principally benefit only the introduced species. Where the translocation or reintroduction involves reinstatement of a genetically similar species, there is little need for discussion. But where new genes are introduced, there is much greater cause for concern (Smith and Holloway, 1989).

Restoration of a site, without translocation or reintroduction, especially within the highly fragmented landscape, mostly benefits mobile eurytopic species. Translocation and reintroduction can supplement the restoration activities and can target and promote the survival of localized, rare or stenotopic species. A knowledge of the species' behavioural ecology and biotope requirements helps establishment.

Translocation and reintroduction involve an understanding of the appropriate life stage to introduce. With butterflies, the adult is the best stage. Roughly 25 eggs, 15 medium-sized larvae or five pupae are equivalent to one adult female (Thomas, 1989). With ladybirds, these proportions are even greater and depend on the host on which the stages have been reared prior to release (Hattingh and Samways, 1991). The most common host, e.g. red scale (*Aonidiella aurantii*) in the case of the coccinellid *Chilocorus, nigritus,* is not as easy to rear as oleander scale *(Aspidiotus nerii)*, and so oleander scale is used in captive breeding. The disadvantage of host change is that on release the ladybird larvae only reluctantly switch host. In contrast, the adults much more readily adjust to a new host.

There has been an increasing tendency in economically rich areas for visually conspicuous insects to be reintroduced as their ranges have become fragmented and reduced. Most introductions have been poorly documented, if at all. Oates and Warren (1990) note 323 recorded attempts at establishment or population reinforcement of Lepidoptera in Britain. These activities involved 43 species, 37 of which were native to Britain. Many of the translocations (42%) never led to establishment of a permanent population. Successful establishment in this case means an introduced colony persisting, unaided, for at least three years in the wild state. Of the remainder, 26% were successful, but the outcome of 32% is unknown. Nearly half (47%) of the British butterfly releases were primarily conducted for conservation purposes, but 29% were releases of surplus breeding stock, 17% were for various amenity purposes and 7% were for scientific experimentation. Although reasons for successful establishment are difficult to assess owing to paucity of information, there was no correlation between numbers of individuals released and success. Establishment was achieved with as few as three or five mated females. However, the main reason for failure appeared to be the unsuitability of the biotope, although susceptibility to environmental change also played a role. In Britain at least, these successful introductions have been highly influential on distribution patterns, with the marsh fritillary *(Eurodryas aurinia)* surviving in more introduced colonies than naturally occurring ones (Figure 11.3).

Oates and Warren (1990) provide some other valuable insights. The conservation of British butterflies would benefit greatly from a more selective establishment programme of only five species. Establishment of other species is either impractical because of lack of suitable biotopes or unnecessary as the species are still distributed widely enough to allow some natural recolonization. The poor colonizing ability of butterflies has meant that many sites that have lost their populations have not regained them. Destruction of habitats and fragmentation of the landscape has had the effect of thinning the distribution pattern of many species. Reintroductions can help solve this problem. The reality is that this approach may be practicable for the well-known and few British butterflies but not so for the vast numbers of yet-to-be identified Amazonian beetles.

11.1.6 Succession

The survival of insects depends largely on the plant composition and the plantscape. As plant communities are successional, either locally (e.g. tree

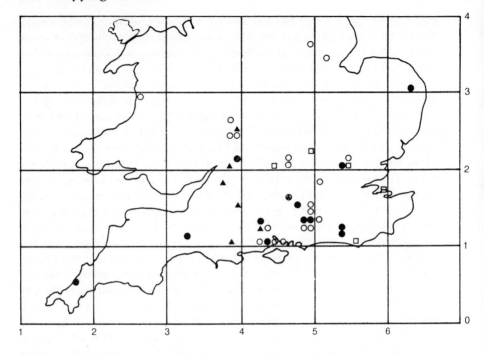

Figure 11.3 The marsh fritillary butterfly (*Eurodryas aurinia*) introductions and population reinforcement in England and Wales. Its distributional range has been kept broad by the several introductions over the years. (From Oates and Warren, 1990.)

fall) or across whole landscapes (e.g. savanna grassland to bush), restoration, reintroduction, translocation and subsequent management must recognize at which stage the plant succession will be halted or modified (Usher and Jefferson, 1991). This depends on the conservation management goals. The restored ecosystems may be halted at a particular stage (e.g. by sheep grazing) or the whole successional process restarted by periodic disturbance (e.g. fire, bush clearance). Research would have to determine the scale and spatial pattern of such disturbances for maintenance of particular insect groups. This is implied by Donnelly and Giliomee's (1985) large-scale study of ants in the Mediterranean-type fynbos ecosystem of the South African Cape, where species richness declined from 29 species 3 months after burning to 20 species 4 years after, and 14 species for the unnaturally long post-burn time of 39 years. Ants are highly competitive, and interspecific relationships probably began to favour certain species over others as the biotopes changed, causing shifts in population distribution.

Restoring a site does not mean that the whole area should be managed in the same way. Studies on the requirements of various grassland Hemiptera (Morris, 1990a,b) and Coleoptera (Morris and Rispin, 1987) suggest that any one site should have a suite of management programmes. The site may have plots with fixed management regimens so that each plot is halted at a particular seral stage. Alternatively, the site could be subjected to rotational management, where all the plots are out of step in their successional position but run to maturity, and then restarted by quite drastic management. This sort of rotational management is especially suitable to harvesting resources such as timber in different patches (plots) at different times (Harris, 1984).

During recreation of sites, it may be appropriate to sow patches of different seed mixtures (Morris, 1990b) to create a patchwork of harlequin biotopes (Denno, 1977). This may also include the planting of specific host plants at a later date, which may be difficult or even impossible to establish at the initial sowing (Morris and Plant, 1983). The conservationists' dilemma, where different species require different seral stages, is partially resolved using fixed plot or rotational management to ensure the continual presence of particular plantscapes, such as tall and short grassland essential for the survival of species populations of different plants and animals (Morris and Rispin, 1987; Rushton, Eyre and Luff, 1990).

11.1.7 Restoration in tropical versus temperate lands

Restoration ecology for conservation is a rapidly developing field in the north-temperate countries. It is taking place much more slowly in the tropics, where human population growth, economic inhibitions, less scientific information and more delicate and complex ecosystems make recovery considerably more formidable. The present Gaia sickness is disseminated primataemia. The causal pathogen is us, by our increasing population level and increasing consumption of natural resources with concurrent pollution (Lovelock, 1991). Population increase is particularly high in the tropics, although natural resources consumption per capita is considerably higher in temperate lands. It is essential that the tropics, which play such an important role in Gaia maintenance and yet are also so fragile, be repaired as soon as possible. This will be slow at first, possibly accelerating with greater recognition of the importance of restored tropical ecosystems (Figure 11.4).

It is now timely and vital that funds be diverted from the north-temperate lands to the tropics to slow the extensive landscape modification and to

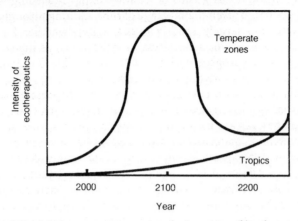

Figure 11.4 Speculative curves suggesting the intensity of landscape rehabilitation and related activities in the future in the temperate zones (upper curve) and tropics (lower curve). (From Soulé, 1989.)

undertake more restoration projects in the spirit of *Caring for the Earth* (IUCN, 1991).

11.1.8 Landscape connectiveness

Biotopes that have a high level of resources enable species to produce a surplus of offspring. These are source biotopes, in contrast to sink biotopes, which do not produce enough offspring to make up for mortality (Pulliam and Danielson, 1991). In a model developed by Danielson (1991), interactions between two species do not result solely from the intrinsic properties of the species involved. One species may cause an increase in the equilibrium population size of a second species, where the high-quality source biotope for this second species is rare relative to the abundance of low-quality sink biotope. This positive effect is facilitation. A negative effect, or inhibition, occurs when the first species reduces the second species when this second species' source biotope is abundant.

Danielson's (1991) point that biotope heterogeneity influences interactions between species, and Webb's (1989) findings that interactions between patches and their matrices also influence species abundances, together with Wood and Samways' (1991) species mobility patterns, all strongly suggest that there is much more to consider when making conservation decisions than simply area size and distances of biotope islands. Additive upon these dynamic features are interactions such as predation, parasitism, mutualism, disease, intraspecific competition and vagaries of weather. Such complexity leaves the conservationist with a bewildering number of management options. Cutting through this great knot, rather than unravelling it, is the most expedient approach given the shortage of time. Cutting through, means rapidly identifying and largely intuitively managing (whether for rarity, endemism, species generation, naturalness or typicalness) as many, and as big, and as connected heterogeneous landscapes as possible. The landscapes that slipped through to degradation can be restored.

At the level of the patch, the corridors may be relatively narrow, measured in a few metres, along which insects may walk or fly. Such corridors and associated refuge patches are important in determining dispersal speed and subsequent distribution patterns. As the corridors become several tens of metres wide, they may also provide for residence, functioning like a long, thin patch (see Burel and Baudry, 1990). The processes and interactions here are primarily at an ecological scale. At an evolutionary scale, where corridors are measured in tens or hundreds of kilometres, the function is also for long-term survival for whole phylogenetic lineages which can disperse and genetically radiate (Figure 11.5).

It is vital to recognize the importance of scale. In terms of human impacts on the land surface, it is also essential to take into account geographical location around the world. Although there may not be vast differences in

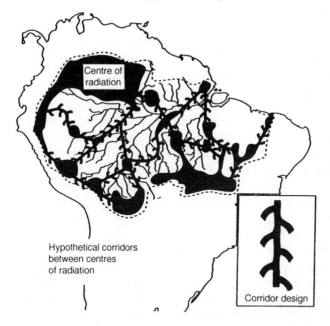

Centre of radiation

Hypothetical corridors between centres of radiation

Corridor design

Figure 11.5 Future corridors of ecological reconstruction between hypothetical centres of radiation, for example in the Amazon Basin, to allow species movements and radiation. Inset is a design for a corridor that maximizes soil and biotope types in small areas. (From Erwin, 1991. Copyright 1991 by the AAAS.)

biological processes, such as insect mobilities, guild patterns etc., from one area to another, there is a great deal of difference between, say, the intensive patchwork of small agricultural fields in the Netherlands compared with the wheatfields of Canada, the pampas of Argentina and the forests of Sulawesi. In the fine mosaic landscapes of north-west Europe, attention has been focused on the finer, ecological time scale, in which hedgerows, small remnant patches and private gardens (Welch, 1990) play an important role in maintaining species with relatively widespread distributions across Europe. In contrast, the research thrust in the tropics has been on a much larger spatial and temporal scale, an evolutionary scale, emphasizing taxonomic distinctiveness. With recognition of species-dynamo areas, it is now timely that restoration ecology apply more of the evolutionary approach to the temperate lands and, much more importantly, introduce more landscape ecology in the tropics.

11.1.9 The future of restoration ecology for insect conservation

The field of restoration ecology for insects is a fertile area of research. In practice, most sites will be restored principally with the plants and perhaps vertebrates in mind. Fully terrestrial as opposed to aquatic sites are the most difficult to restore. This is especially so if the disturbance has changed the soil composition, if there is seepage at mine sites, or if there is modification of the local topography, albeit apparently minor. Aquatic biotopes offer more predictiveness, particularly in hindsight after long-term studies (Murphy and Weiss, 1988; Moore, 1991a). But time is desperately short for biodiversity decisions to be made on such admirable studies as the 27-year odonatological study of Moore (1991a). A good underlying study of the species present, and the ecology of an area (see Spellerberg, 1991) adjacent to a major disturbance, or a study of a truly pristine patch, such as a fragment of prairie, and to restore it to its former condition is to some extent feasible (Jordan, 1988). This assumes that the socioeconomic context is also taken into account (Todd, 1988).

11.1.10 Insect reserves

Local populations of both temperate and tropical insects die out and recolonization occurs when dispersal is adequate. However, repopulation does not always occur. This poses a dilemma, because setting aside a single reserve for a single species rests on inherent risks, biological and climatic. Even with dragonflies, which often fairly readily colonize new sites, it is necessary to have a collection of sites to allow recolonization when local extinction prevails (Moore, 1991a). Nevertheless, setting aside certain small reserves for specific highly range-restricted, high-profile species does seem to be feasible in some cases. Lycaenid butterflies have received particular attention in areas and countries of high economic activity. New (1991) reviews some of the examples, with all case histories having the common denominator of being high in research requirements to describe the autecology in detail, demanding intensive management input, being demographically susceptible to changes in the biotope and to adversity in weather, and being expensive. Such reserves in no way guarantee survival of the species, even less so considering global climate changes.

Setting aside reserves for community conservation, rather than for single species, is a more realistic prospect, assuming the specified landscape carries a sufficiently large population or a metapopulation (Opdam, 1990). Such a land area should ideally have biotope and topographic variety for the populations to spatially move in accordance with plantscape and weather changes. They also need refuge from biotic interactions such as disease and interspecific competition.

Such biodiversity reserves must integrate with the surrounding matrix. This boundary is a skin whose condition indicates the health of the system (Schonewald-Cox, 1988). Nothing short of maintenance of the reserve as intact as possible, with appropriate management, is the best option. Naturally, the human context also prevails and the boundary in this respect plays an important role (McCarthy, 1991).

11.2 BREEDING PROGRAMMES AND PRESERVATION TECHNOLOGY

11.2.1 Breeding for conservation

Captive breeding of insects has several facets. One area, which only indirectly concerns conservation, is the breeding of utility insects such as biocontrol agents and other living material, such as silkworms (*Bombyx mori*), and cochineal (*Dactylopius opuntiae*) (section 8.6). As the ecological significance of insects is not generally recognized by the public, captive breeding has done much in terms of conservation awareness of smaller animals. There are about 38–40 butterfly houses in Britain, with their popularity gaining elsewhere. Captive breeding allows study of the basic biology and perfection of rearing techniques (Lees, 1989). Captive breeding has the potential for developing techniques for continuously maintaining cultures of rare and scientifically useful species. In Papua New Guinea, birdwing butterfly breeding has moved into the realm of farming for export of papered specimens for the collectors' market (Parsons, 1984). This is quite a different concept from the rearing of rare species for reintroduction into the wild.

Captive breeding for conservation certainly has some potential, particularly if carried out as part of a captive breeding organizations' commercial enterprise. Aside from the commercial value of rare specimens, the breeding stock is available for introduction into the field to supplement existing populations. The stock, however, must be genetically fit and similar to the native stock, and a good knowledge of the natural habitat is required. Bearing in mind the severe fragmentation of many landscapes and the ease with which many butterflies can be reared, Morton (1983) presents a cogent case for the establishment of a captive breeding institute.

Captive breeding is unquestionably a last resort. Many species, even well-known abundant species, can be extremely difficult to rear continuously in the laboratory. The cassava hornworm (*Erinnyis ello*) (a hawkmoth) simply will not oviposit when spatially confined. Likewise, no method is yet known for continuously rearing most Anisoptera in captivity. There are many other practical problems, such as supplying suitable host plant or host animal material, or providing an adequate alternative. Without the appropriate nutrition, individuals are often diminutive and weak, and ill-adapted to open-air conditions. This is the case with ladybirds (e.g. *Chilocorus nigritus*). Also, within the insectary, light and temperature regimens must be varied. Many decticine bush crickets (e.g. *Platycleis intermedia*) do not sing or mate and therefore produce no offspring under uniform lighting and temperature conditions. Also, during development, some species physiologically and behaviourally lock onto a food plant that is provided in captivity, and then cannot recognize or develop on their wild foodplant once released. Also, after the climatically mild and possible disease-free, predator-free and parasitoid-free, conditions of the rearing cages, the wild environment is often too adverse for the offspring to survive.

Genetic restrictions such as genetic drift and inadvertent selection for survival under insectary conditions, rather than wild conditions, can occur. In biological control circles this may be overcome by introducing wild strains from the natural biotope. In conservation terms, this becomes an increasing paradox as fewer and fewer wild individuals become available, with the insectary also increasingly unable to produce fit individuals.

These points suggest that captive breeding is certainly not the answer or even a substitute for conservation of individuals in the natural landscape. Nevertheless, it provides a vital role for rapid boosting of population numbers for reintroduction and as a final means of saving a species should its extirpation be imminent. Where it is used as a method for restocking wild biotopes, sufficient numbers of individuals need to be released. Thomas (1989) suggests a minimum of 30 female butterflies, otherwise it is difficult to determine whether any failure is due to chance influences or to inverse density-dependent factors. Yet the interesting finding of Oates and Warren (1990) that re-establishment of butterflies has been achieved with as few as three to five mated females is encouraging.

Monitoring after release is vital, and logistically not difficult at a specific biotope. Widespread dispersal after release may well indicate a lack of tendency to establish as is the case with some biocontrol agents.

In terms of candidate species for captive breeding and release towards their conservation, Lees (1989) lists several British macro-Lepidoptera, and tropical Lepidoptera, Orthoptera, Phasmatodea, Dermaptera, Coleoptera and Diptera. These are principally high-profile species because of their glamour, size and rarity, which provide sufficient motivation and funds for their care and survival.

Captive breeding of insects clearly offers some potential for certain threatened species. This is aside from its commercial and heuristic value. In reality however, bearing in mind the vast number of insects, it is a form of exclusive conservation to a few relatively subjectively important species, and is not of the same magnitude of importance as biotope, landscape and global conservation.

11.2.2 Preservation technology

Captive breeding is the first step in preservation technology. It involves individuals of selected species continuing their life processes within the physical confines of a rearing cage. In this environment there is often physiological and genetic deterioration. Stopping the life processes, not through death, but by genetic freezing technology, and also the changing of the species character by biotechnology and methodologies, is now coming about.

Different organisms, whether viruses, bacteria, fungi, plants, amphibia, mammals or insects require different approaches and techniques. There are several hurdles to be overcome in preservation technology and genetic engineering, some general and some specific to insects.

Firstly, there are enormous moral issues to face. Has wild nature and evolution really come to an end as a result of global pollution and genetic manipulation? Or are the preservation technologists simply defying time but not life? Genetic engineering has many and far-reaching consequences that may make the early warnings on pesticide usage seem insignificant. We know not what we do. Any community ecologist or biological control practitioner will suggest caution, caution and more caution. We are already seeing the tip of the iceberg risk with modifying baculoviruses for lepidopteran pest control (section 8.4).

The second hurdle is that we know so little of any genome, even the human one. Even for the geneticists' best friend, *Drosophila melanogaster*, with its possibly 10 000 genes and about 14 million base pairs, the maps of even small parts of its genome are poor. If we multiply by the number of insect species, say 10 million-fold, the production of even a few genome maps would make just a tiny dent in this vast genetic world. In the future, it may not be an impossibility, but the rate of genome map production will not match the current rate of species loss.

This leads to the third hurdle, which is the extremely high cost of genome mapping. Time is short and there is the additional dilemma of choosing which of the insects', or other organisms', genomes to map. The economic and pragmatic incentives are not there. This rebounds on to the bioempathetic approach for saving many of earth's numerous species. Survival of earth's biotic variety depends on us respecting its future simply for its own sake and not necessarily for ours.

In addition, how long can the spark of life be maintained in a suspended state? Recent advances in molecular palaeontology have allowed us to reconstruct aspects of the biology of a Cretaceous beetle, but this does not bring that beetle back to life.

Unquestionably, preservation technology has a vast research potential. Methodologies such as cryopreservation or alternatives need to be developed. Choice of which life stage to preserve will also be important, whether gamete, egg, pupa or adult. Also, for lodging in genetic libraries, an adequate sample size of individuals that truly represent the species also needs to be selected. Maintenance of future eradicated pests in suspended animation, as in the case of the smallpox virus, is another consideration. 'Eradication' is not the term used in pest management circles, with 'suppression' generally being more appropriate. Nevertheless, with future technologies, eradication of both medically and agriculturally important pests may become a reality.

These preservation techniques and approaches will inevitably focus on significant vertebrates and angiosperms. Insects, however, may be useful experimental subjects, if not preservation targets themselves. Of greater significance, especially to insect and ecosystem processes in general, will be the development of techniques for holding whole communities in a suspended life state. Future restoration ecology may well involve reseeding landscapes with these resuscitated communities.

11.2.3 Overcoming the taxonomic impediment

At the present time one of the greatest restraints on describing species is the dearth of expertise and funds for taxonomy. This problem is now being partially tackled, at least from the financial angle, by sponsor-a-species schemes. It involves naming the type of a species after a sponsor, using the sponsor's surname, company name, or any other name as the specific or even, where feasible, the generic name. The idea was spawned after seven Costa Rican parasitic hymenoptera raised $300 000 towards protection of their biotopes by being named after seven executives who sponsored the scheme.

Meanwhile our task of describing the species on earth must continue, with taxonomists of the future likely to be image analyser specialists and computer programmers who can scan perhaps hundreds of species per day. The technology, in addition to describing and sorting the specimens, would compose phylogenetic trees based on the scanned morphologies. Coupled with genetic mapping, there would be a powerful technology for identifying geographical areas for naturalness, as species dynamos and as endemically curious areas. Over the decades, there is also likely to be the means of recreating these species-groups and even resiting them once global climate change has stabilized.

11.3 SUSTAINABLE CONSERVATION

11.3.1 Comparative approaches

Restoration ecology principally benefits common, eurytopic species, while captive breeding's conservation approach tends to include only specially selected species. Preservation ecology may eventually benefit both ends of the spectrum. Insect conservationists working on landscape design and management point out that for the rarer and more demanding species better management of existing biotopes is the prime approach, at least in Europe. Elsewhere, preservation, with or without management, of various biotopes in various landscapes is the main approach to conserving insect variety. Linking with botanists is vital, as plants are stationary, form the plantscape and are major determinants of landscape character. Plants are also relatively well-known, particularly in comparison with insects.

Inventories of species in conserved and non-conserved areas are vital in determining the status of many species. In tropical and sub-tropical areas the task is formidable. Recognizing phylogenetic quality areas and species dynamos is an important first step. Use of selected taxonomic indicator groups for assessing the quality of those sites, big and small, over time, can help monitor population and species survival in the face of landscape fragmentation and global climate change.

11.3.2 The human factor

There are various areas of interface between humans and insects. Apart from an awareness of a tiny minority of species as pests, there is generally an enormous impediment to the perception of the importance of insects in ecosystem processes. Any global conservation plan must take in the needs for human survival as well as the needs of biota, as outlined with succinctness and with realistic strategic planning in *Caring for the Earth* (IUCN, 1991). The importance of the interface between the human and insect factors, mediated by plants or by direct interaction in the case of medically important insects, has not been fully realized. This is not entomological chauvinism but a simple statement of the fact that man as a species is impacting most upon the earth, while insects are the biomass majority, their abundance being 30 times that of humans per area in the USA. Man is damaging the diversity structure of the biosphere, insects being the major component of that diversity.

Sustainable use of resources is a main tenet of *Caring for the Earth*. With insects, this applies to their plantscape rather than themselves. Despite the fact that insects are taxonomically varied and high in numbers of individuals, they are exasperatingly difficult to harvest. With the exception of selection of some products such as honey and cochineal dye, and some localized and occasional harvesting of some individual Isoptera, Lepidoptera, Orthoptera and Coleoptera, insects have little to offer. The small size, chitinous exoskeleton and great population variability makes harvesting them an unattractive enterprise. It is the ecosystem services, as predators, parasitoids, nutrient cyclers and pollinators, for which they must be conserved. To achieve this, their biotopes and landscapes must be used sustainably, after first being carefully identified and conserved. The catch-all approach of landscape conservation is the most realistic.

People must benefit if insect conservation is to be maximized. One cannot sell bioethics to the starving, but one can illustrate the value of insects. The task of the insect conservation biologist is to relay their research findings to the biodiversity preservation spokespeople. Those research findings may be on a single high-profile temperate insect, the structure and functioning of the rain forest or desert, the last remaining island outpost of an unusual endemic, or the fragility of the fauna of a particular cave, or even models of the impact of global warming. All this information should be relayed to both the politicians and managers of today, and to our children for future management. Without the flutter or buzz of the insects assuring us that the ecosystems are healthy, there would indeed be a great loneliness of spirit.

11.4 SUMMARY

As so many species, ecosystems and landscapes are under threat, we cannot afford pessimism. Although the doomsday alerts have been essential for action, we now need positive constructivism based on sound restoration-ecology science. Man is under less threat from landscape change than the other species on earth. Possibly 98% of earth's diversity is in terrestrial ecosystems, most of which is vested in insects. Conservation of as many of these systems in as many landscapes as possible is the optimal approach. This conservation additionally involves recreating biotopes and restoring landscapes.

Planting of trees on the easily eroded, easily compacted and readily oxidized soils of the tropics is urgently required. Research in the temperate lands has shown that sowing of grass and wildflower seed mixtures, plus later introduction of selected plants, can be highly beneficial to insects.

Restoration of sites tends to encourage mobile, widespread and eurytopic insect species. Localized and less mobile species require introduction, assuming the plantscape is suitable for receiving them. To obtain this knowledge, detailed autecological studies are needed. This emphasizes that restoration ecology is not an immediate panacea for conserving qualitative biodiversity. The survival of specific stenotopic threatened species may depend on their being captively bred and then released into suitable sites. Translocation of individuals from one location to another may also determine whether a species survives or not. Establishment, whether from reintroduction or translocation, depends on a thorough knowledge of the insects' biologies.

After a site has been restored and even restocked with species, part or whole of it may be halted by vegetation management at a particular successional stage to suit the conservation goals. Rotational management of the various seral stages at different adjacent subsites can be used to maximize insect variety.

Restoration ecology in the temperate lands is continuing rapidly, but in the tropics it has had a slow start. Recognition of the enormous value of restoring ecosystems in the tropics for maintaining the earth's landscapes, biotic diversity and climate is vitally urgent. Restoration involves introducing patches, preferably large ones, which ideally should connect with each other and link with remnant and resource patches. This linking needs to occur at various scales of size and time. At the size of the immediate landscape, linkages allow mobility and the survival of insect populations, although there may also be some risk through creating passages for predators and pathogens. At the larger scale of whole wilderness areas, massive movement corridors operate at the evolutionary level as well as at smaller, ecological scales. These radiation corridors permit movement and phylogenetic radiation, thus contributing to promoting future biodiversity.

Breeding insects in captivity for conservation has given insects a high profile in the eyes of the public. This emphasizes indirectly that insects are creatures of note and essential in terrestrial ecosystem processes. Captive breeding has various logistical and genetic problems, but has nevertheless proved extremely valuable for many high-profile species. It is principally exclusive to those species, and not an answer for general biodiversity conservation.

Preservation technology, the holding of genetic and physiological processes in suspended animation, has enormous potential. It must be clearly distinguished from genetic engineering, in which species are modified, and which holds great environmental risks. Mapping genomes of insects presents a hugely expensive technological hurdle, at a time when many species are becoming extinct. Preserving species in genetic libraries for future examination and reintroduction has to contend with the problem of how long the life spark can be held and rereleased along with its genome. Other technological advances may well include the preservation of whole communities for long-term restoration projects once the global climate has been stabilized.

Image analysers and future genetic freezing techniques will enable many species to be rapidly described, catalogued and compared, phylogenetically, one with another.

The most urgent, morally sound and sensible alternative is to preserve as many biotopes and landscapes as intact as possible. As the human factor must also be built into any contemporary conservation strategy, sustainable utilization must be considered. Insects, in view of their small size, high chitin content, great population variability and harvesting difficulties, generally are not particularly valuable as a direct resource for humans. Their value lies in the role they play in ecosystem processes. The landscapes in which they occur must be conserved and utilized sustainably. Insects should be preserved *per se*, so that their activities can maintain ecosystems and also benefit man. Besides, insects, as with other biota, are our earthly companions. Without them we would indeed suffer from a great loneliness of spirit.

References

Acheta domestica, M.E.S. (1849)*Episodes of Insect Life,* Reeve, Benham & Reeve, London. (This was published anonymously under the pen-name of 'the hearth cricket'. The author however, was Mrs M.B. Budgen.)

Ackery, P.R. and Vane-Wright, R.I. (1984) *Milkweed Butterflies,* British Museum (Natural History), London.

Adis, J. (1990) Thirty million arthropod species – too many or too few? *Journal of Tropical Ecology,* **6,** 115–118.

Alcock, J. (1987) Leks and hilltopping in insects. *Journal of Natural History,* **21,** 319–328.

Allen, A.A. (1951–64) The Coleoptera of a suburban garden. *Entomologists' Records and Journal of Variation,* **63,** 61–64; 187–190; **64,** 61–63, 92–93; **65,** 225–231; **68,** 215–222; **71,** 16–20, 39–44; **76,** 237–242, 261–264.

Altieri, M.A. (1989) Agroecology: A new research and development paradigm for world agriculture. *Agriculture, Ecosystems and Environment,* **27,** 37–46.

Altmann, M. (1992) 'Biopesticides' turning into new pests. *Trends in Ecology and Evolution,* **7,** 65.

Andersen, A.N. (1990) The use of ant communities to evaluate change in Australian terrestrial ecosystems: a review and recipe. *Proceedings of the Ecological Society of Australia,* **16,** 347–357.

Andow, D.A. (1991a) Vegetational diversity and arthropod population response. *Annual Review of Entomology,* **36,** 561–586.

Andow, D.A. (1991b) Yield loss to arthropods in vegetationally diverse agroecosystems. *Environmental Entomology,* **20,** 1228–1235.

Andow, D.A. and Hidaka, K. (1989) Experimental natural history of sustainable agriculture: syndromes of production. *Agriculture, Ecosystems and Environment,* **27,** 447–462.

Annecke, D.P. and Moran, V.C. (1977) Critical reviews of biological pest control in South Africa. 1. The Karoo caterpillar, *Loxostege frustalis* Zeller (Lepidoptera: Pyralidae), *Journal of the Entomological Society of Southern Africa,* **40,** 127–145.

Anonymous (1983) When biocontrol goes wrong. *Biocontrol News and Information,* **4,** 1.

Anonymous (1986) The Assisi Event. *The New Road,* Issue No. 1, Winter 1986–87, World Wide Fund for Nature, Gland, Switzerland.

Anonymous (1990) Judaism, Jain Faith, Baha'i Faith, Sikhism,Christianity, Buddhism, Islam, Hinduism, *The New Road,* Issue No. 16, Oct.–Dec. 1990, World Wide Fund for Nature, Gland, Switzerland.

Askew, R.R. (1988) *The Dragonflies of Europe*, Harley Books, Colchester, England.

Atkinson, I. (1989) Introduced animals and extinctions, in *Conservation for the Twenty-First Century* (eds D. Western and M. Pearl), Oxford University Press, Oxford, pp. 54–69.

Aubert, B. and Quilici, S. (1983) Nouvel équilibre biologique observé à la Réunion sur les populations de psyllides aprés l'introduction et l'établissement d'hymenoptéres chalcidiens. *Fruits*, **38**, 771–780.

Austin, M.P. and Margules, C.R. (1986) Assessing representativeness, in *Wildlife Conservation Evaluation* (ed. M.B. Usher), Chapman & Hall, London, pp. 45–67.

Bedford, E.C.G. (1949) Report of the Plant Pathologist. Report of the Department of Agriculture, Bermuda, pp. 11–19.

Bei-Bienko, G.Y. (1970) Orthopteroid insects (Orthopteroidea) of the National Park areas near Kursk and their significance as indices of the local landscape. *Zhurnal Obschchei Biologii*, **31**, 30–46. (In Russian, translated by the British Library, Lending Division.)

Beirne, B.P. (1985) Avoidable obstacles to colonization in classical biological control of insects. *Canadian Journal of Zoology*, **63**, 743–747.

Bennett, K.D. (1986) The rate of spread and population increase of forest trees during the postglacial. *Philosophical Transactions of the Royal Society of London, B*, **314**, 533–540.

Benton, M.J. (1986) The evolutionary significance of mass extinctions. *Trends in Ecology and Evolution*, **1**, 127–130.

Berry, R.J. (1971) Conservation aspects of the genetical constitution of populations, in *The Scientific Management of Animal and Plant Communities for Conservation* (eds E. Duffey and A.S. Watt), Blackwell Scientific Publications, Oxford, pp. 177–206.

den Boer, P.J. (1990) The survival value of dispersal in terrestrial arthropods. *Biological Conservation*, **54**, 175–192.

Bond, W.J. and Breytenbach, G.J. (1985) Ants, rodents and seed predation in Proteaceae. *South African Journal of Zoology*, **20**, 150–154.

Bond, W.J. and Slingsby, P. (1984) Collapse of an ant–plant mutualism: the Argentine ant (*Iridomyrmex humilis*) and myrmecochorous Proteaceae. *Ecology*, **65**, 1031–1037.

Booth, M. and Allen, M.M. (1990) Butterfly garden design, in *Butterfly Gardening* (created by The Xerces Society in association with The Smithsonian Institution), Sierra Club Books, San Francisco, pp. 69–93.

Borror, D.J. and White, R.E. (1970) *A Field Guide to the Insects of America North of Mexico*, Houghton Mifflin, Boston.

Boyd, H.P. and Marucci, P.E. (1979) Arthropods of the Pine Barrens, in *Pine Barrens: Ecosystem and Landscape* (ed. R.T.T. Forman), Academic Press, New York, pp. 505–517.

Bradley, J.D. and Mere, R.M. (1966) Natural history of the garden of Buckingham Palace. Further records and observations 1964–1965. Lepidoptera. *Proceedings of the South London Entomological and Natural History Society*, **1966**, 15–17.

Bradshaw, A.D., Goode, D.A. and Thorp, E.H.P. (eds) (1986) *Ecology and Design in Landscape*, Blackwell Scientific Publications, Oxford.

Brakefield, P.M. (1987) Industrial melanism: do we have the answers? *Trends in Ecology and Evolution*, **2**, 117–122.

Brakefield, P.M. (1989) The variance on genetic diversity among subpopulations is more sensitive to founder effects or bottlenecks than is the mean: a case study, in *Evolutionary Biology of Unstable Transient Populations* (ed. A. Fontdevila), Springer-Verlag, Berlin, pp. 145–161.

Brakefield, P.M. (1991) Genetics and the conservation of invertebrates, in *The Scientific Management of Temperate Communities for Conservation* (eds I.F. Spellerberg, F.B. Goldsmith and M.G. Morris), Blackwell Scientific Publications, Oxford, pp. 45–79.

Braker, E. (1991) Natural history of a neotropical gap-inhabiting grasshopper. *Biotropica*, **23**, 41–50.

Briggs, J.C. (1991) Global species diversity. *Journal of Natural History*, **25**, 1403–1406.

Brower, L.P. and Malcolm, S.B. (1989) Endangered phenomena. *Wings*, **14**, 3–10.

Brown, J.H. and Davidson, D.W. (1977) Competition between seed-eating rodents and ants in desert ecosystems. *Science*, **196**, 880–882.

Brown, K.S. (1982) Paleoecology and regional patterns of evolution in neotropical forest butterflies, in *Biological Diversification in the Tropics* (ed. G.T. Prance), Columbia University Press, New York, pp. 255–308.

Brown, K.S. (1991) Conservation of neotropical environments: insects as indicators, in *The Conservation of Insects and their Habitats* (eds N.M. Collins and J.A. Thomas), 15th Symposium of the Royal Entomological Society of London, Academic Press, New York, pp. 350–404

Brown, V.K. (1991) The effects of changes in habitat structure during succession in terrestrial communities, in *Habitat Structure: The Physical Arrangement of Objects in Space* (eds S.S. Bell, E.D. McCoy and H.R. Mushinsky), Chapman & Hall, London, pp. 141–168.

Brown, V.K., Hendrix, S.D. and Dingle, H. (1987) Plants and insects in early old-field succession: comparison of an English site and an American site. *Biological Journal of the Linnean Society*, **31**, 59–74.

Brown, V.K. and Southwood, T.R.E. (1983) Trophic diversity, niche breadth and generation times of exopterygote insects in a secondary succession. *Oecologia*, **56**, 220–225.

Bryant, E.H., Meffert, L.M. and McCommas, S.A. (1990) Fitness rebound in serially bottlenecked populations of the housefly. *American Naturalist,* **136,** 542–549.

Buckley, G.P. (ed.) (1989) *Biological Habitat Reconstruction,* Belhaven, London.

Bunce, R.G.H. and Howard, D.C. (eds) (1990) *Species Dispersal in Agricultural Habitats,* Belhaven, London.

Burel, F. (1992) Effect of landscape structure and dynamics on species diversity in hedgerow networks. *Landscape Ecology,* **6,** 161–174.

Burel, F. and Baudry, J. (1990) Hedgerow networks as habitats for forest species: implications for colonising abandoned agricultural land, in *Species Dispersal in Agricultural Habitats* (eds R.G.H. Bunce and D.C. Howard), Belhaven, London, pp. 238–255.

Burkey, T.V. (1989) Extinction in nature reserves: the effect of fragmentation and the importance of migration between reserve fragments. *Oikos,* **55,** 75–81.

Busby, J.R. (1991) BIOCLIM – A bioclimate analysis and prediction system, in *Nature Conservation* (eds C.R. Margules and M.P. Austin), CSIRO, Canberra, pp. 64–67.

Callicott, J.B. (1986) On the intrinsic value of non-human species, in *The Preservation of Species* (ed. B.G. Norton), Princeton University Press, Princeton, New Jersey, pp. 138–172.

Caltagirone, L.E. (1981) Landmark examples in classical biological control. *Annual Review of Entomology,* **26,** 213–232.

Caltagirone, L.E. and Huffaker, C.B. (1980) Benefits and risks of using predators and parasites for controlling pests, in *Environmental Protection and Biological Forms of Control of Pest Organisms* (eds B. Lundholm and M. Stackerud), *Ecological Bulletin* Vol. 31, Stockholm, pp. 103–109.

Campbell, A.J. and Tainton, M.T. (1981) Effects of fires on the invertebrate fauna of soil and litter of a eucalypt forest, in *Fire and the Australian Biota* (eds A.M. Gill, R.H. Groves and J.R. Noble), Australian Academy of Science, Canberra, pp. 215–241.

Carroll, C.R., Vandermeer, J.J. and Rossett, P.M. (eds) (1990) *Agroecology,* McGraw-Hill, New York.

Carson, R. (1963) *Silent Spring,* Hamish Hamilton, London.

Cassagnau, P. (1959) Contribution à la connaissance du genre *Tetracanthella* Schott. *Mémoires du Muséu d'Histoire Naturelle,* Paris (Zool.), **16,** 201–260.

Chapin III, F.S., Schulze, E.–D. and Mooney, H.A. (1992) Biodiversity and ecosystem processes. *Trends in Ecology and Evolution,* **7,** 107–108.

Chapman, P. (1987) Extinction faces the world's cave creatures. *IUCN Bulletin,* **18,** No. 7–9, 13–14.

Cheng, L. (ed.) (1976) *Marine Insects,* North-Holland, Amsterdam.

Cherrill, A.J. and Brown, V.K. (1990a) The life cycle and distribution of the wart-biter *Decticus verrucivorus* (L.) (Orthoptera: Tettigoniidae) in a chalk grassland in southern England. *Biological Conservation,* 53, 125–143.

Cherrill, A.J. and Brown, V.K. (1990b) The habitat requirements of adults of the wart-biter *Decticus verrucivorus* (L.) (Orthoptera: Tettigoniidae) in southern England. *Biological Conservation,* 53, 145–157.

Chown, S.L. (1990) Possible effects of Quaternary climatic change on the composition of insect communities of the South Indian Ocean Province Islands. *South African Journal of Science,* 86, 386–391.

CIBC (1982–83) Oil palm pollination, in *Commonwealth Institute of Biological Control, Annual Report, 1982–83.* Commonwealth Agricultural Bureaux, Slough, England, p.12.

Cilliers, C.J. and Neser, S. (1991) Biological control of *Lantana camara* (Verbenaceae) in South Africa. *Agriculture, Ecosystems and Environment,* 37, 57–75.

Claridge, M.F. and Singhrao, J.S. (1978) Diversity and altitudinal distribution of grasshoppers (Acridoidea) on a Mediterranean mountain. *Journal of Biogeography,* 5, 239–250.

Clark, T.E. (1991) Dragonflies as habitat indicators of the Sabie River in the Kruger National Park. Unpublished M.Sc thesis, University of Natal.

Clark, T.E. and Samways, M.J. (1994) Ecological landscaping for conservation of macro-arthropod diversity in a southern hemisphere (South African) urban botanic garden, in *Habitat Creation and Wildlife Conservation in Post-industrial and Urban Habitats* (eds J. Rieley and S. Page), Packard, UK.

Cobham, R. and Rowe, J. (1986) Evaluating wildlife of agricultural environments: an aid to conservation, in *Wildlife Conservation Evaluation* (ed. M.B. Usher), Chapman & Hall, London, pp. 223–246.

Cohn, J.P. (1989) Gauging the biological impacts of the greenhouse effect. *BioScience,* 39, 143–146.

Collier R. (1986) *The Conservation of the Chequered Skipper in Britain.* Report No. 16 of the Interpretative Services Branch. Nature Conservancy Council, Peterborough, UK.

Collier, R.H., Finch, S., Phelps, K. and Thompson, A.R. (1991) Possible impact of global warming on cabbage root fly *(Delia radicum)* activity in the UK. *Annals of Applied Biology,* 118, 261–271.

Collins, N.M. (1987) *Butterfly Houses in Britain: the Conservation Implications,* IUCN, Cambridge, UK, and Gland, Switzerland.

Collins, N.M. (General ed.) (1990) *The Last Rain Forests,* IUCN, Gland, Switzerland, and Mitchell Beazley, London.

Collins, N.M. (1991) Insect conservation – pointers for the future. *Antenna,* 15, 73–78.

Collins, N.M. and Morris, M.G. (1985) *Threatened Swallowtail Butterflies of the World.* IUCN, Gland, Switzerland, and Cambridge, UK.

Collins, N.M. and Thomas, J.A. (eds) (1991) *The Conservation of Insects and their Habitats*, 15th Symposium of the Royal Entomological Society of London, Academic Press, London.

Collins, N.M. and Wells, S.M. (1987) *Invertebrates in Need of Special Protection in Europe*, European Committee for the Conservation of Nature and Natural Reserves, Strasbourg.

Connell, J. and Sousa, W.P. (1983) On the evidence needed to judge ecological stability or persistence. *American Naturalist*, **121**, 789–824.

Connor, E.F. (1986) The role of Pleistocene forest refugia in the evolution and biogeography of tropical biotas. *Trends in Ecology and Evolution*, **1**, 165–168.

Conservation International (1990) *The Rain Forest Imperative*, Conservation International, Washington DC.

Coope, G.R. (1975) Mid-Weichselian climatic changes in western Europe, re-interpreted from coleopteran assemblages, in *Quaternary Studies* (eds R.P. Suggate and M.M. Cresswell), Royal Society of New Zealand, Wellington, pp. 101–108.

Cottrell, C.B. (1985) The absence of co-evolutionary associations with Capensis Floral Element plants in the larval/plant relationships of Southwestern Cape butterflies, in *Species and Speciation* (ed. E.S. Vrba), Transvaal Museum Monograph No.4, Transvaal Museum, Pretoria, pp. 115–124.

Cousins, S.H. (1991) Species diversity measurement: choosing the right index. *Trends in Ecology and Evolution*, **6**, 190–192.

Cox, C.B. and Moore, P.D. (1985) *Biogeography: An Ecological and Evolutionary Approach*, 4th edn, Blackwell Scientific Publications, Oxford.

Crossley, D.A., Coleman, D.C. and Hendrix, P.F. (1989) The importance of the fauna in agricultural soils: research approaches and perspectives. *Agriculture, Ecosystems and Environment*, **27**, 47–55.

Culver, D.C. (1986) Cave faunas, in *Conservation Biology: The Science of Scarcity and Diversity* (ed. M.E. Soulé), Sinauer Associates, Sunderland, Massachusetts, pp. 427–443.

Danielson, B.J. (1991) Communities in a landscape: the influence of habitat heterogeneity on the interactions between species. *The American Naturalist*, **138**, 1105–1120.

Darlington, A. (1981) *Ecology of Walls*, Heinemann, London.

Dasmann, R.F. (1959) *Environmental Conservation*, Wiley, New York.

Dasmann, R.F. (1984) *Environmental Conservation*, 5th edn, Wiley, New York.

Davidson, D.W. (1977) Species diversity and community organization in desert seed-eating ants. *Ecology*, **58**, 711–724.

Davies, R.E. and Koch, R.H. (1991) All the observed universe has contributed to life. *Philosophical Transactions of the Royal Society, London, B*, **334**, 391–403.

Davis, B.N.K. (1978) Urbanisation and the diversity of insects, in *Diversity of Insect Faunas* (eds L.A. Mound and N. Waloff), Blackwell Scientific Publications, Oxford, pp. 126–138.

Davis, B.N.K. (1989) Habitat creation for butterflies on a landfill site. *The Entomologist,* **108,** 109–122.

Davis, S.D., Droop, S.J.M., Gregerson, P., Henson, L., Leon, C.J., Villa-Lobos, J.L., Synge, H. and Zantovska, J. (1986) *Plants in Danger: What do we Know?* IUCN, Gland, Switzerland.

Dawkins, M.S. (1989) Attitudes to animals, in *The Fragile Environment* (eds L. Friday and R. Laskey), Cambridge University Press, Cambridge, UK, pp. 41–60.

Day, D. (1989) *The Encyclopedia of Vanished Species* (Revised Edition), Mclaran, Hong Kong.

De Angelis, D.L. and Waterhouse, J.C. (1987) Equilibrium and non-equilibrium concepts in ecological models. *Ecological Monographs,* **57,** 1–21.

DeBach, P. (1974) *Biological Control by Natural Enemies,* Cambridge University Press, Cambridge.

De Kock, A.E. (1990) Interactions between the introduced Argentine ant, *Iridomyrmex humilis* Mayr, and two indigenous fynbos ant species. *Journal of the Entomological Society of Southern Africa,* **53,** 107–108.

De Kock, A.E. and Giliomee, J.H. (1989) A survey of the Argentine ant *Iridomyrmex humilis* (Mayr) (Hymenoptera, Formicidae) in South African fynbos. *Journal of the Entomological Society of Southern Africa,* **52,** 157–164.

Dempster, J.P. (1971) Some effects of grazing on the population ecology of the Cinnabar Moth (*Tyria jacobaeae* L.), in *The Scientific Management of Animal and Plant Communities for Conservation* (eds E. Duffey and A.S. Watt), Blackwell Scientific Publications, Oxford, pp. 517–526.

Dempster, J.P. (1975) *Animal Population Ecology,* Academic Press, London.

Dempster, J.P. (1989) Insect introductions: natural dispersal and population persistence in insects. *The Entomologist,* **108,** 5–13.

Dempster, J.P. (1991) Fragmentation, isolation and mobility of insect populations, in *The Conservation of Insects and their Habitats* (eds N.M. Collins and J.A. Thomas), 15th Symposium of the Royal Entomological Society of London, pp. 143–153.

Dempster, J.P. and Pollard, E. (1986) Spatial heterogeneity, stochasticity and detection of density dependence in animal populations. *Oikos,* **46,** 413–423.

Dennis, R.L.H. (1977) *The British Butterflies, Their Origins and Distribution,* Classey, Faringdon, UK.

Dennis, R.L.H. and Shreeve, T.G. (1991) Climatic change and the British butterfly fauna: opportunities and constraints. *Biological Conservation,* **55,** 1–16.

Denno, R.F. (1977) Comparisons of the assemblages of sap-feeding insects (Homoptera–Hemiptera) inhabiting two structurally different salt marsh grasses in the genus *Spartina*. *Environmental Entomology*, **6**, 359–372.

Denno, R.F. and Roderick, G.K. (1991) Influence of patch size, vegetation texture, and host plant architecture on the diversity, abundance, and life history styles of sap-feeding herbivores, in *Habitat Structure: The Physical Arrangement of Objects in Space* (eds S.S. Bell, E.D. McCoy and H.R. Mushinsky), Chapman & Hall, London, pp. 169–196.

Descimon, H. (1986) Origins of lepidopteran faunas in the high tropical Andes, in *High Altitude Tropical Biogeography* (eds F. Vuilleumier and M. Monasterio), Oxford University Press, New York, pp. 533–556.

Desender, K. and Turin, H. (1989) Loss of habitats and changes in the composition of the ground and tiger beetle fauna in four west European countries since 1950 (Coleoptera: Carabidae, Cicindelidae), *Biological Conservation*, **48**, 277–294.

Diamond, J.M. (1975) The island dilemma: lessons of modern biogeographic studies for the design of natural reserves. *Biological Conservation*, **7**, 129–146.

Digby, P.G.N. and Kempton, R.A. (1987) *Multivariate Analysis of Ecological Communities*, Chapman & Hall, London.

Disney, R.H.L. (1986) Assessments using invertebrates: posing the problem, in *Wildlife Conservation Evaluation* (ed.M.B. Usher), Chapman & Hall, London, pp. 271–293.

Dobzhansky, T. (1950) Evolution in the tropics. *American Scientist*, **38**, 209–221.

Dobzhansky, T. and Pavan, C. (1950) Local and seasonal variations in relative frequencies of species of *Drosophila* in Brazil. *Journal of Animal Ecology*, **19**, 1–14.

Donnelly, D. and Giliomee, J.H. (1985) Community structure of epigaeic ants in a pine plantation and in newly burnt fynbos. *Journal of the Entomological Society of southern Africa*, **48**, 259–265.

Dover, J.W. (1991) The conservation of insects on arable farmland, in *The Conservation of Insects and their Habitats* (eds N.M. Collins and J.A. Thomas), 15th Symposium of the Royal Entomological Society of London, pp. 293–318.

Duffey, E. (1968) Ecological studies on the large copper butterfly *Lycaena dispar* Haw. *batavus* Obth. at Woodwalton Fen National Nature Reserve, Huntingdonshire. *Journal of Applied Ecology*, **5**, 69–96.

Duffey, E. (1971) The management of Woodwalton Fen: a multidisciplinary approach, in *The Scientific Management of Animal and Plant Communities for Conservation* (eds E. Duffey and A.S. Watt), Blackwell Scientific Publications, Oxford, pp. 581–597.

Duffey, E. (1974) *Nature Reserves and Wildlife*, Heinemann, London.

Duffey, E. (1977) The re-establishment of the large copper butterfly *Lycaena dispar batava* Obth. on Woodwalton Fen National Nature Reserve, Cambridgeshire, England, 1969–1973. *Biological Conservation*, **12**, 143–158.

Duffey, E. and Watt, A.S. (eds) (1971) *The Scientific Management of Animal and Plant Communities for Conservation*, Blackwell Scientific Publications, Oxford.

Duncan, J. (1840) *Introduction to Entomology*, Lizars, Edinburgh and London.

Durrell, L. (1986) *State of the Ark*, The Bodley Head, London.

Eardley, C.D. (1989) Diversity and endemism of southern African bees. *Plant Protection News,* No. 18, December 1989, pp. 1–2.

Eastop, V.F. (1978) Diversity of Sternorrhyncha within major climatic zones, in *Diversity of Insect Faunas* (eds L.A. Mound and N. Waloff), 9th Symposium of the Royal Entomological Society, Blackwell Scientific Publications, Oxford, pp. 71–88.

Ehrenfeld, D.W. (1970) *Biological Conservation*, Holt, Rinehart & Winston, New York.

Ehrlich, P.R. and Murphy, D.D. (1987) Conservation lessons from long-term studies of checkerspot butterflies. *Conservation Biology*, **1**, 122–131.

Ehrlich, P.R. and Raven, P.H. (1964) Butterflies and plants: a study in co-evolution. *Evolution*, **18**, 586–608.

Elliott, J.M. (1991) Aquatic insects as target organisms for the study of effects of projected climate change in the British Isles. *Freshwater Forum*, **1**, 195–203.

Elton, C.S. (1927) *Animal Ecology*, Sidgwick & Jackson, London.

Elton, C.S. (1958) *The Ecology of Invasions by Animals and Plants*, Methuen, London.

Elton, C.S. (1966) *The Pattern of Animal Communities*, Methuen, London.

Erhardt, A. and Thomas, J.A. (1991) Lepidoptera as indicators of change in the semi-natural grasslands of lowland and upland Europe, in *The Conservation of Insects and their Habitats* (eds N.M. Collins and J.A. Thomas), 15th Symposium of the Royal Entomological Society of London, Academic Press, London, pp. 213–236.

Erwin, T.L. (1982) Tropical forests: their richness in Coleoptera and other Arthropod species. *Coleopterists' Bulletin,* **36**, 74–75.

Erwin, T.L. (1988) The tropical forest canopy: the heart of biotic diversity, in *Biodiversity* (ed. E.O. Wilson), National Academy Press, Washington DC, pp. 123–129.

Erwin, T.L. (1991) An evolutionary basis for conservation strategies. *Science*, **253**, 750–752.

Fearnside, P.M. (1986) *Human Carrying Capacity of the Brazilian Rain Forest*, Columbia University Press, New York.

Ferrar, A.A. (1989) The role of Red Data Books in conserving biodiversity, in *Biotic Diversity in Southern Africa: Concepts and Conservation* (ed. B.J. Huntley), Oxford University Press, Cape Town, pp. 136–147.

Fischer, A.G. (1960) Latitudinal variations in organic diversity. *Evolution*, **14**, 64–81.

Fisher, J., Simon, N. and Vincent, J. (1969) *The Red Book: Wildlife in Danger*, Collins, London.

Ford, E.B. (1975) *Ecological Genetics*, 4th edn, Chapman and Hall, London.

Forman, R.T.T. and Baudry, J. (1984) Hedgerows and hedgerow networks in landscape ecology. *Environmental Management*, **8**, 495–510.

Forman, R.T.T. and Godron, M. (1986) *Landscape Ecology*, Wiley, New York.

Forney, K.A. and Gilpin, M.E. (1989) Spatial structure and population extinction: a study with *Drosophila* flies. *Conservation Biology*, **3**, 45–51.

Foster, G.N. (1991) Conserving insects of aquatic and wetland habitats, with special reference to beetles, in *The Conservation of Insects and their Habitats* (eds N.M. Collins and J.A. Thomas), 15th Symposium of the Royal Entomological Society of London, Academic Press, London, pp. 237–262.

Foster, G.N., Foster, A.P., Eyre, M.D. and Bilton, D.T. (1990) Classification of water beetle assemblages in arable fenland and ranking of sites in relation to conservation value. *Freshwater Biology*, **22**, 343–510.

Franz, J.M. (1964) Dispersion and natural enemy action. *Annals of Applied Biology*, **53**, 510–515.

Fry, G.L.A. (1991) Conservation in agricultural ecosystems, in *The Scientific Management of Temperate Communities for Conservation* (eds I.F. Spellerberg, F.B. Goldsmith and M.G. Morris), Blackwell Scientific Publications, Oxford, pp. 415–443.

Fry, R. and Lonsdale, D. (1991) *Habitat Conservation for Insects – A Neglected Green Issue*, The Amateur Entomologists' Society, Feltham, Middlesex, England.

Furley, P.A., Newey, W.W., Kirby, R.P. and Mc G. Hotson, J. (1983) *Geography of the Biosphere*, Butterworths, London.

Gadagkar, R., Chandrashekara, K. and Nair, P. (1990) Insect species diversity in the tropics: sampling methods and a case study. *Journal of the Bombay Natural History Society*, **87**, 337–353.

Galecka, B. (1991) Distribution and role of coccinellids in an agricultural landscape, in *Behaviour and Impact of Aphidophaga* (eds L. Polgár, R.J. Chambers, A.F.G. Dixon and I. Hodek), Academic Publishing, The Hague, pp. 137–141.

Gandar, M.V. (1982) The dynamics and trophic ecology of grasshoppers (Acridoidea) in a South African savanna. *Oecologia (Berlin)*, **54**, 370–378.

Gaston, K.J. (1991) The magnitude of global insect species richness. *Conservation Biology,* 5, 283–296.

Gaston, K.J. (1992) Regional numbers of insect and plant species. *Functional Ecology,* 6, 243–247.

Gaston, K.J. and Lawton, J.H. (1990) Effects of scale and habitat on the relationship between regional distribution and local abundance. *Oikos,* 58, 329–335.

Gauch, H.G. (1982) *Multivariate Analysis in Community Ecology,* Cambridge University Press, Cambridge.

Gibbs Russell, G.E. (1985) Analysis of the size and composition of the southern African flora. *Bothalia,* 15, 613–629.

Giberson, D.J., Rosenberg, D.M. and Wiens, A.P. (1991) Changes in abundance of burrowing mayflies in Southern Indian Lake: Lessons for environmental monitoring. *Ambio,* 20, 139–142.

Gilbert, L.F. (1980) Food web organization and the conservation of neotropical diversity, in *Conservation Biology: An Evolutionary–ecological Perspective* (eds M.E. Soulé and B.A. Wilcox), Sinauer Associates, Massachusetts, pp. 11–34.

Gilbert, O.L. (1989) *The Ecology of Urban Habitats,* Chapman & Hall, London.

Gilbertson, D.D., Kent, M. and Pyatt, F.B. (1985) *Practical Ecology for Geography and Biology, Survey, Mapping and Data Analysis,* Unwin Hyman, London.

Giliomee, J.H (1992) Factors disturbing insects in a fynbos reserve. *Proceedings of th VIth International Conference on Mediterranean Climate Ecosystems, Maleme, Crete, Greece (23–27 September 1991)* pp. 133-139.

Gill, A.M., Groves, R.H. and Noble, I.R. (eds) (1981) *Fire and the Australian Biota,* Australian Academy of Science, Canberra.

Gilpin, M.E. (1987) Experimental community assembly: competition, community structure and the order of species introductions, in *Restoration Ecology: A Synthetic Approach to Ecological Research.* (eds W.R. Jordan, III., M.E. Gilpin and J.D. Aber), Cambridge University Press, Cambridge, pp. 151–161.

Gilpin, M.E. and Soulé, M.E. (1986) Minimum viable populations: processes of species extinction, in *Conservation Biology: The Science of Scarcity and Diversity* (ed. M.E. Soulé), Sinauer Associates, Sunderland, Massachusetts, pp. 19–34.

Goldsmith, O. (1866) *A History of Earth and Animated Nature,* Blackie, Glasgow.

Goode, D.A. and Smart, P.J. (1986) Designing for wildlife, in *Ecology and Design in Landscape* (eds A.D. Bradshaw, D.A. Goode and E. Thorp), Blackwell Scientific Publications, Oxford, pp. 219–235.

Goudie, A. (1989) The changing human impact, in *The Fragile Environment* (eds L. Friday and R. Laskey), Cambridge University Press, Cambridge, pp. 1–21.

Graedel, T.E. and Crutzen, P.J. (1989) The changing atmosphere. *Scientific American*, **261**, 28–36.

Grassle, J.F. and Maciolek, N.J. (1992) Deep-sea species richness: regional and local diversity estimates from quantitative bottom samples. *The American Naturalist*, **139**, 313–341.

Gray, A.J. (1991) Management of coastal communities, in *The Scientific Management of Temperate Communities for Conservation* (eds I.F. Spellerberg, F.B. Goldsmith and M.G. Morris), Blackwell Scientific Publications, Oxford, pp.227–243.

Greenslade, P.J.M. (1983) Adversity selection and the habitat templet. *American Naturalist*, **122**, 352–365.

Greenslade, P. and New, T.R. (1991) Australia: Conservation of a continental insect fauna, in *The Conservation of Insects and their Habitats* (eds N.M. Collins and J.A. Thomas), 15th Symposium of the Royal Entomological Society of London, Academic Press, London. pp. 33–70.

Greenwood, S.R. (1987) The role of insects in tropical forest food webs. *Ambio*, **16**, 267–270.

Groombridge, B. (Ed.) (1992) *Global Biodiversity. Status of the Earth's Living Resources.* Chapman & Hall, London.

Guéguen-Genest, M.C. and Guéguen, A. (1987) Effet du pâturage ovin sur la dynamique de population du criquet de Sibérie *Gomphocerus sibiricus* Finot Orthoptère, acrididae dans une formation pâturée d'altitude. *Comptes rendus Hebdomadaires des Séances de l'Académie des Sciences, Paris. Série III*, **304**, 443–447.

Guilford, T. (1990) The secrets of aposematism: unlearned responses to specific colours and patterns. *Trends in Ecology and Evolution*, **5**, 323–329.

Haes, E.C.M. (1987) *Mogoplistes squamiger,* Scaly Cricket, in *British Red Data Books 2. Insects.* (ed. D.B. Shirt), Nature Conservancy Council, Peterborough, UK, pp. 51–52.

Hall, A.V., de Winter, B., Fourie, S.P. and Arnold, T.M. (1984) Threatened plants in southern Africa. *Biological Conservation*, **28**, 5–20.

Hammock, B.D., Bonning, B.C., Possee, R.D., Hanzlik, T.N. and Maeda, S. (1990) Expression and effects of the juvenile hormone esterase in a baculovirus virus. *Nature*, **344**, 458–461.

Hammond, C.O. (1983) *The Dragonflies of Great Britain and Ireland*, 2nd edn, revised by R. Merritt, Harley, Colchester, UK.

Hammond, P.M. (1992) Species inventory, in *Global Biodiversity: Status of the Earth's Living Resources* (ed. B. Groombridge). Chapman & Hall, London. pp. 17–39.

Hanski, I. and Tiainen, J. (1988) Populations and communities in changing agroecosystems in Finland. *Ecological Bulletins*, **39**, 159–168.

Harris, L.D. (1984) *The Fragmented Forest*, The University of Chicago Press, Chicago and London.

Harris, L.D. (1988) Edge effects and conservation of biotic diversity. *Conservation Biology*, **2**, 330–332.

Harris, L.D. and Gallagher, P.B. (1989) New initiatives for wildlife conservation: the need for movement corridors, in *Defense of Wildlife: Preserving Communities and Corridors* (ed. G. Mackintosh), Defenders of Wildlife, Washington DC, pp. 11–24.

Harris, L.D. and Scheck, J. (1991) From implication to applications: the dispersal corridor principle applied to the conservation of biological diversity, in *Nature Conservation 2: The Role of Corridors* (eds D.A. Saunders and R.J. Hobbs), Surrey Beatty, Chipping Norton, New South Wales, Australia, pp. 189–220.

Harrison, P. (1990) Too much life on Earth? *New Scientist*, **126**, 28–29.

Harrison, S., Murphy, D.D. and Ehrlich, P.R. (1988) Distribution of the bay checkerspot butterfly, *Euphydryas editha bayensis*: evidence for a metapopulation model. *The American Naturalist*, **132**, 360–382.

Hassell, M.P. (1985) Insect natural enemies as regulating factors. *Journal of Animal Ecology*, **54**, 323–334.

Hattingh, V. and Samways, M.J. (1991) Determination of the most effective method for field establishment of biocontrol agents of the genus *Chilocorus* (Coleoptera: Coccinellidae), *Bulletin of Entomological Research*, **81**, 169–174.

Heath, J. (1981) *Threatened Rhapalocera (Butterflies) in Europe*, Council of Europe (Nature and Environment Series No. 23), Strasbourg.

Henning, S.F. and Henning, G.A. (1985) South Africa's endangered butterflies. *Quagga*, **10**, 16–17.

Henning, S.F. and Henning, G.A. (1989) *South African Red Data Book – Butterflies*. Foundation for Research Development, Council for Scientific and Industrial Research, Pretoria.

Hewitt, G.M. (1990) Limited introgression through orthopteran hybrid zones. *Boletin de Sanidad Vegetal Plagas* (Fuera de Serie), **20**, 329–339.

Hill, L. and Michaelis, F.B. (1988) *Conservation of Insects and Related Wildlife*, Report on the Australian National Parks and Wildlife Service, Canberra, Occasional Paper No. 13, 40 pp.

Hill, M.O. (1973) Diversity and evenness: a unifying notation and its consequences. *Ecology*, **54**, 427–431.

Hochberg, M.E. and Waage, J.K (1991) Control engineering. *Nature*, **352**, 16–17.

Hodek, I. (1973) *Biology of Coccinellidae*, Junk, The Hague.

Hodkinson, I.D. and Casson, D. (1991) A lesser predilection for bugs: Hemiptera (Insecta) diversity in tropical rain forests. *Biological Journal of the Linnaean Society*, **43**, 101–109.

Hoffmann, A.A. and Parsons, P.A. (1989) An integrated approach to environmental stresses and life-history variation: Desiccation tolerance in *Drosophila. Biological Journal of the Linnean Society,* **37**, 117–136.

Hoffmann, M.T. and Cowling, R.M. (1990) Vegetation change in the semi-arid eastern Karoo over the last 200 years: an expanding Karoo – fact or fiction? *South African Journal of Science,* **86**, 286–294.

Holdgate, M.W. (1960) The fauna of the mid-Atlantic islands. *Proceedings of the Royal Society, B,* **152**, 550–567.

Holloway, J.D. (1986) Origins of lepidopteran faunas in high mountains of the Indo-Australian tropics, in *High Altitude Tropical Biogeography* (eds F.Vuilleumier and M.Monasterio), Oxford University Press, New York, pp. 533–556.

Horn, D.J. (1991) Potential impact of *Coccinella septempunctata* on endangered Lycaenidae in Northwestern Ohio, in *Behaviour and Impact of Aphidophaga* (eds L. Polgár, R.J. Chambers, A.F.G. Dixon and I. Hodek), Academic Publishing, The Hague, pp. 159–162.

Houck, M.A., Clark, J.B., Peterson, K.R. and Kidwell, M.G. (1991) Possible horizontal transfer of *Drosophila* genes by the mite *Proctolaelaps regalis. Science,* **253**, 1125–1129.

House, G.J. and Brust, G.E. (1989) Ecology of low-input, no-tillage agroecosystems. *Agriculture, Ecosystems and Environment,* **27**, 331–345.

Howarth, F.G. (1983) Classical biological control: panacea or Pandora's box. *Proceedings of the Hawaiian Entomological Society,* **24**, 239–244.

Howarth, F.G. (1991) Environmental impacts of classical biological control. *Annual Review of Entomology,* **36**, 485–509.

Howarth, F.G. and Ramsay, G. (1991) The conservation of island insects and their habitats, in *The Conservation of Insects and their Habitats* (eds N.M. Collins and J.A. Thomas) 15th Symposium of the Royal Entomological Society of London, Academic Press, London. pp. 71–119.

Hoyle, F. (1948) in Weber, R. (1986) *Dialogues with Scientists and Sages: The Search for Unity,* Routledge & Kegan Paul, London.

Huey, R.B. (1991) Physiological consequences of habitat selection. *The American Naturalist,* **137**, S91–S115.

Huntley, B.J. (1988) Conserving and monitoring biotic diversity: some South African examples, in *Biodiversity,* (ed. E.O. Wilson), National Academy Press, Washington DC, pp. 248–260.

Huntley, B.J., Siegfried, R. and Sunter, C. (1989) *South African Environments into the 21st Century,* Human and Rousseau Tafelberg, Cape Town.

Irish, J. (1989) Biospeleology of Dragon's Breath Cave, South West Africa/Namibia, *Proceedings of the 7th Entomological Congress, Entomological Society of Southern Africa,* Pietermaritzburg 10–13 July 1989, p. 73.

IUCN (1980) *World Conservation Strategy,* IUCN/UNEP/WWF, Gland, Switzerland.

IUCN (1987) Special Report: Invertebrates – no safety in numbers. *IUCN Bulletin*, **18**, 1–20.

IUCN (1990) *The 1990 IUCN Red List of Threatened Animals*, IUCN, Gland, Switzerland, and Cambridge, UK

IUCN (1991) *Caring for the Earth: A Strategy for Sustainable Living*, IUCN/UNEP/WWF, Gland, Switzerland.

Jackson, J.K. and Resh, V.H. (1989) Sequential decision plans, benthic macroinvertebrates, and biological monitoring programmes. *Environmental Management*, **13**, 455–468.

Jacobs, M. (1988) *The Tropical Rain Forest. A First Encounter*, Springer-Verlag, Berlin.

Jaenike, J. and Holt, R.D. (1991) Genetic variation for habitat preference: evidence and explanations. *The American Naturalist*, **137**, S67–S90.

Jago, N.D. (1973) The genesis and nature of tropical forest and savanna grasshopper faunas, with special reference to Africa, in *Tropical Forest Ecosystems in Africa and South America: A Comparative Review* (eds B.J. Meggers, E.S. Ayensu and D. Duckworth), Smithsonian Institution Press, Washington DC, pp. 187–196.

Jägomagi, J., Külvik, M., Mander, U. and Jacuchno, V. (1988) The structural functional role of ecotones in the landscape. *Ecology (CSSR)*, **7**, 81–94.

Janzen, D.H. (1975) *Ecology of Plants in the Tropics*, Edward Arnold, London.

Janzen, D.H. (1988) Tropical dry forests: The most endangered major tropical ecosystem, in *Biodiversity* (ed. E.O.Wilson), National Academy Press, Washington DC, pp. 130–137.

Janzen, D.H. (ed.) (1989) *Costa Rican Natural History,* University of Chicago Press, Chicago.

Johnsen, P. (1987) The status of the South African Acridoidea s.l. (Orthoptera: Caelifera), in *Evolutionary Biology of Orthopteroid Insects* (ed. B.M. Baccetti), Ellis Horwood, Chichester, England, pp. 293–295.

Johnson, D.L. (1989a) Spatial autocorrelation, spatial modelling, and improvements in grasshopper survey methodology. *Canadian Entomologist*, **121**, 579–588.

Johnson, D.L. (1989b) Spatial analysis of the relationship of grasshopper outbreaks to soil classification, in *Estimation and Analysis of Insect Populations. Lecture Notes in Statistics*, **55**, 347–359.

Johnson, D.L. and Worobec, A. (1988) Spatial and temporal computer analysis of insects and weather: grasshoppers and rainfall in Alberta. *Memoirs of the Entomological Society of Canada*, **146**, 33–48.

Jones, S. (1991) Hedgerows, in *Habitat Conservation for Insects – A Neglected Green Issue* (eds R. Fry and D. Lonsdale), The Amateur Entomologists' Society, Middlesex, England.

Jongman, R.H.G., ter Braak, C.J.F. and van Tongeren, O.F.R. (1987) *Data*

Analysis in Community and Landscape Ecology, Pudoc, Wageningen.

Jordan, W.R. III (1988) Ecological restoration: reflections on a half-century of experience at the University of Wisconsin–Madison Arboretum, in *Biodiversity* (ed. E.O.Wilson), National Academy Press, Washington DC, pp. 311–316.

Jordan, W.R. III, Gilpin, M.E. and Aber, J.D. (eds) (1987) *Restoration Ecology: A Synthetic Approach to Ecological Research,* Cambridge University Press, Cambridge.

Joyce, C. (1989) Dying to get on the list. *New Scientist,* **127,** 42–47.

Julien, M.H. (ed.) *Biological Control of Weeds: A World Catalogue of Agents and their Target Weeds,* Commonwealth Agricultural Bureaux, Wallingford, UK.

Keals, N. and Majer, J.D. (1991) The conservation status of ant communities along the Wubin–Perenjori corridor, in *Nature Conservation 2. The Role of the Corridors* (eds D.A. Saunders and R.J. Hobbs), Surrey Beatty, Chipping Norton, New South Wales, Australia, pp. 387–393.

Kellogg, W.W. and Schware, R. (1981) *Climate Change and Society: Consequences of Increasing Atmospheric Carbon Dioxide,* Westview Press, Boulder, Colorado.

Kemp, J.C. and Barrett, G.W. (1989) Spatial patterning: Impact of uncultivated corridors on arthropod populations within soybean agroecosystems. *Ecology,* **70,** 114–128.

Kettlewell, H.B.D. (1973) *The Evolution of Melanism,* Clarendon Press, Oxford.

Kingdon, J. (1990) *Island Africa: The Evolution of Africa's Rare Animals and Plants,* Collins, London.

Kirk, W.D.J. (1991) The size relationship between insects and their hosts. *Ecological Entomology,* **16,** 351–359.

Kluge, R.L. (1991) Biological control of crofton weed, *Ageratina adenophora* (Asteraceae), in South Africa. *Agriculture, Ecosystems and Environment,* **37,** 187–191.

Kluge, R.L. and Caldwell, P.M. (1992) Microsporidian diseases and biological weed control agents: to release or not to release? *Biocontrol News and Information,* **13,** 43N-47N.

Kornberg, H. and Williamson, M.H. (eds) (1986) Quantitative aspects of the ecology of biological invasions. *Philosophical Transactions of the Royal Society, London, B,* **314,** 501–742.

Kudrna, O. (1986) *Butterflies of Europe: Aspects of the Conservation of Butterflies in Europe,* AULA-Verlag, Wiesbaden.

Kusnezov, N. (1957) Numbers of species of ants in faunae of different latitudes. *Evolution,* **11,** 298–299.

Ladle, M. (1991) Running waters: a conservationist's nightmare, in *The Scientific Management of Temperate Communities for Conservation* (eds

I.F. Spellerberg, F.B. Goldsmith and M.G. Morris), Blackwell Scientific Publications, Oxford, pp. 383–393.

Laurance, W.F. (1991) Edge effects in tropical forest fragments: application of a model for the design of nature reserves. *Biological Conservation*, **57**, 205–219.

Laurance, W.F. and Yensen, E. (1991) Predicting the impacts of edge effects in fragmented habitats. *Biological Conservation*, **55**, 77–92.

Lawton, J.H. (1983) Plant architecture and diversity of phytophagous insects. *Annual Review of Entomology*, **28**, 23–39.

Lawton, J.H. and Strong, D.R. Jr (1981) Community patterns and competition in folivorous insects. *The American Naturalist*, **118**, 317–338.

Lees, D. (1989) Practical considerations and techniques in the captive breeding of insects for conservation purposes. *The Entomologist*, **108**, 77–96.

Leopold, A. (1933) *Game Management*, Charles Scribner & Sons, New York.

Leopold, A. (1949) *A Sand County Almanac and Sketches Here and There*, Oxford University Press, Oxford.

Leston, D. (1973) The ant mosaic – tropical tree crops and the limiting of pests and diseases. *Pest Articles and News Summaries*, **19**, 311–341.

Lewis, T. (1969a) The distribution of flying insects near a low hedgerow. *Journal of Applied Ecology*, **6**, 443–452.

Lewis, T. (1969b) The diversity of the insect fauna in a hedgerow and neighbouring field. *Journal of Applied Ecology*, **6**, 443–452.

Lisansky, S., Robinson, A. and Coombs, J. (1991) *The UK Green Growers' Guide*, CPL Press, Newbury, UK.

Lockwood, J.A. (1987) The moral standing of insects and the ethics of extinction. *The Florida Entomologist*, **70**, 70–89.

Lockwood, J.A. and De Brey, L.D. (1990) A solution for the sudden and unexplained extinction of the Rocky Mountain Grasshopper (Orthoptera: Acrididae), *Environmental Entomology*, **19**, 1194–1205.

Louw, G.N. and Seely, M.K. (1982) *Ecology of Desert Organisms*, Longman, London.

Lovejoy, T.E., Bierregaard, R.O., Rylands, A.B., Malcolm, J.R., Quintela, C.E., Harper, L.H., Brown, K.S. Jr, Powell, A.H., Powe., G.V.N., Schubart, H.O.R. and Hays, M.B. (1986) Edge and other effects of isolation on Amazon Forest fragments, in *Conservation Biology: The Science of Scarcity and Diversity* (ed. M.E. Soulé), Sinauer, Sunderland, Massachusetts, pp. 257–285.

Lovelock, J.E. (1979) *Gaia. A New Look at Life on Earth*, Oxford University Press, Oxford.

Lovelock, J.E. (1988) *The Ages of Gaia: A Biography of our Living Earth*, Oxford University Press, Oxford.

Lovelock, J.E. (1991) *Gaia: The Practical Science of Planetary Medicine*, Gaia Books, London.

Ludwig, J.A. and Reynolds, J.F. (1988) *Statistical Ecology: A Primer on Methods and Computing*, Wiley, New York.

Lyman, F. (1990) *The Greenhouse Trap*, Beacon, Boston.

MacArthur, R.H. (1972) *Geographical Ecology: Patterns in the Distribution of Species*, Harper & Row, New York.

MacArthur, R.H. and Wilson, E.O. (1967) *The Theory of Island Biogeography*, Princeton University Press, Princeton.

Macdonald, I.A.W., Kruger, F.J. and Ferrar, A.A. (eds) (1986) *The Ecology and Management of Biological Invasions in Southern Africa*, Oxford University Press, Oxford.

McCarthy, J. (1991) Sites of Special Scientific Interest: The British experience. *Natural Areas Journal,* **11**, 108–113.

McCoy, E.D. and Bell, S.S. (1991) Habitat structure: The evolution and diversification of a complex topic, in *Habitat Structure: The Physical Arrangement of Objects in Space* (eds S.S. Bell, E.D. McCoy and H.R. Mushinsky), Chapman & Hall, London, pp. 3–27.

McCoy, E.D., Bell, S.S. and Mushinsky, H.R. (1991) Habitat structure: synthesis and perspectives, in *Habitat Structure: The Physical Arrangement of Objects in Space* (eds S.S. Bell, E.D. McCoy and H.R. Mushinsky), Chapman & Hall, London.

McDonnell, M.J. and Pickett, S.T.A. (1990) Ecosystem structure and function along urban–rural gradients: an unexploited opportunity for ecology. *Ecology,* **71**, 1232–1237.

MacGarvin, M., Lawton, J.H. and Heads, P.A. (1986) The herbivorous insect communities of open and woodland bracken: observations, experiments and habitat manipulations. *Oikos,* **47**, 135–148.

McGeoch, M.A. and Samways, M.J. (1991) Dragonflies (Odonata: Anisoptera) and the thermal landscape: implications for their conservation. *Odonatologica,* **20**, 303–320.

McKechnie, S.W., Ehrlich, P.R. and White, R.R. (1975) Population genetics of *Euphydryas* butterflies. I. Genetic variation and the neutrality hypothesis. *Genetics,* **81**, 571–594.

McKibben, B. (1990) *The End of Nature*, Viking, London.

McNeely, J.A., Miller, K.R., Reid, W.V., Mittermeier, R.A. and Werner, T.B. (1990) *Conserving the World's Biological Diversity.* IUCN, WRI, CI, WWF-US, the World Bank, Gland, Switzerland, and Washington DC.

Mace, G.M. and Lande, R. (1991) Assessing extinction threats: toward a reevaluation of IUCN threatened species categories. *Conservation Biology,* **5**, 148–157.

Mader, H.J. (1984) Animal habitat isolation by roads and agricultural fields. *Biological Conservation,* **29**, 81–96.

Magurran, A.E. (1988) *Ecological Diversity and its Measurement,* Croom Helm, London.

Malcolm, S.B. (1990) Mimicry: status of a classical evolutionary paradigm. *Trends in Ecology and Evolution,* 5, 57–62.

Malthus, T.R. (1798) *An Essay on the Principle of Population as it affects the Future Improvements of Society,* London (reprinted by Macmillan, New York).

Mani, M.S. (1968) *Ecology and Biogeography of High Altitude Insects,* Junk, The Hague.

Manly, B.F.J. (1986) *Multivariate Statistical Analysis: A Primer,* Chapman & Hall, London.

Margules, C.R. and Austin, M.P. (eds) (1991) *Nature Conservation: Cost Effective Biological Surveys and Data Analysis,* CSIRO, Canberra.

Marshall, J.A. and Haes, E.C.M. (1988) *Grasshoppers and Allied Insects of Great Britain and Ireland,* Harley, Colchester, UK.

May, R.M. (1978) The dynamics and diversity of insect faunas, in *Diversity of Insect Faunas* (eds L.A. Mound and N. Waloff), 9th Symposium of the Royal Entomological Society, Blackwell Scientific Publications, Oxford, pp. 188–204.

May, R.M. (1986) How many species are there? *Nature,* 324, 514–515.

May, R.M (1988) How many species are there on earth? *Science,* 241, 1441–1449.

May, R.M. (1989) How many species?, in *The Fragile Environment,* (eds L. Friday and R. Laskey), Cambridge University Press, Cambridge, pp. 61–81.

May, R.M (1990a) Taxonomy as destiny. *Nature,* 347, 129–130.

May, R.M. (1990b) How many species? *Philosophical Transactions of the Royal Society,* B330, 293–304.

Mellanby, K. (1967) *Pesticides and Pollution,* Collins, London.

Mielke, H.W. (1989) *Patterns of Life: Biogeography of a Changing World,* Unwin Hyman, London.

Mikkola, K. (1991) The conservation of insects and their habitats in northern and eastern Europe, in *The Conservation of Insects and their Habitats* (eds N.M. Collins and J.A. Thomas), 15th Symposium of the Royal Entomological Society of London, Academic Press, London, pp. 109–119.

Mikkola, K. and Albrecht, A. (1986) Radioactivity in Finnish night-flying moths (Lepidoptera) after the Chernobyl accident. *Notulae Entomologicae,* 66, 153–157.

Mitchell, R.G. and Martin, R.E. (1980) Fire and insects in pine culture of the Pacific Northwest. *Proceedings of the Conference on Fire Forest Meteorology,* 6, 182–190.

Mitter, C., Farrell, B. and Futuyma D.J. (1991) Phylogenetic studies of insect–plant interactions: insights into the genesis of diversity. *Trends in Ecology and Evolution,* 6, 290–293.

Mittermeier, R.A. (1988) Primate diversity and the tropical forest: Case studies from Brazil and Madagascar and the importance of the megadiversity countries, in *Biodiversity* (ed. E.O. Wilson), National Academy Press, Washington, DC. pp. 145–154.

Mooney, H.A. and Drake, J.A. (eds) (1986) *The Ecology of Biological Invasions of North America and Hawaii*, Springer-Verlag, New York.

Moore, N.W. (1987a) *The Bird of Time: The Science and Politics of Nature Conservation*, Cambridge University Press, Cambridge.

Moore, N.W. (1987b) A most exclusive preserve. *New Scientist*, 115, 64.

Moore, N.W. (1991a) The development of dragonfly communities and the consequences of territorial behaviour: a 27-year study on small ponds at Woodwalton Fen, Cambridgeshire, United Kingdom. *Odonatologica*, 20, 203–231.

Moore, N.W. (1991b) Observe extinction or conserve diversity?, in *The Conservation of Insects and their Habitats* (eds N.M. Collins and J.A. Thomas), 15th Symposium of the Royal Entomological Society of London, Academic Press, London, pp. 1–8.

Moore, N.W. and Gagné, W.C. (1982) *Megalagrion pacificum* (McLachlan): A preliminary study of the conservation requirements of an endangered species. Reports of the Odonata Specialist Group, Species Survival Commission, IUCN. No.3, Societas Internationalis Odonatologica, Utrecht, 5 pp.

Mopper, S., Maschinski, J., Cobb, N. and Whitham, T.G. (1991) A new look at habitat structure: Consequences of herbivore-modified plant architecture, in *Habitat Structure: The Physical Arrangement of Objects in Space* (eds S.S. Bell, E.D. McCoy, and H.R. Mushinsky), Chapman & Hall, London, pp. 250–280.

Moran, V.C. (1980) Interactions between phytophagous insects and their *Opuntia* hosts. *Ecological Entomology*, 5, 153–164.

Moran, V.C. and Southwood, T.R.E. (1982) The guild composition of arthropod communities in trees. *Journal of Animal Ecology*, 51, 289–306.

Morris, M.G. (1971) The management of grassland for the conservation of invertebrate animals, in *The Scientific Management of Animal and Plant Communities for Conservation* (eds E.Duffey and A.S. Watt), Blackwell Scientific Publications, Oxford, pp. 527–552.

Morris, M.G. (1990a) The Hemiptera of two sown calcareous grasslands. III. Comparisons with the Auchenorrhyncha faunas of other grasslands. *Journal of Applied Ecology*, 27, 394–409.

Morris, M.G. (1990b), The Hemiptera of two calcareous grasslands. II. Differences between treatments. *Journal of Applied Ecology* 27, 379–393.

Morris, M.G. (1991) The management of reserves and protected areas, in *The Scientific Management of Temperate Communities for Conservation* (eds I.F. Spellerberg, F.B. Goldsmith and M.G. Morris), Blackwell Scientific Publications, Oxford, pp. 323–347.

Morris, M.G. and Plant, R. (1983) Responses of grassland invertebrates to management by cutting. V. Changes in Hemiptera following cessation of management. *Journal of Applied Ecology*, **20**, 157–177.

Morris, M.G. and Rispin, W.E. (1987) Abundance and diversity of the coleopterous fauna of a calcareous grassland under different cutting régimes. *Journal of Applied Ecology*, **24**, 451–465.

Morris, M.G., Collins, N.M., Vane-Wright, R.I. and Waage, J. (1991) The utilization and value of non-domesticated insects, in *The Conservation of Insects and their Habitats* (eds N.M. Collins and J.A. Thomas), 15th Symposium of the Royal Entomological Society of London, Academic Press, London, pp. 319–347.

Morse, D.R., Stork, N.E. and Lawton, J.H. (1988) Species number, species abundance and body length relationships of arboreal beetles in Bornean lowland rain forest. *Ecological Entomology*, **13**, 25–37.

Morton, A.C. (1983) Butterfly conservation – the need for a Captive Breeding Institute. *Biological Conservation*, **25**, 19–33.

Murphy, D.D.(1988) Challenges to biological diversity in urban areas, in *Biodiversity* (ed. E.O. Wilson), National Academy Press, Washington DC, pp. 71–76.

Murphy, D.D. (1989) Conservation and confusion: wrong species, wrong scale, wrong conclusions. *Conservation Biology*, **3**, 82–87.

Murphy, D.D. (1990) Conservation biology and scientific method. *Conservation Biology*, **4**, 203–204.

Murphy, D.D., Freas, K.E. and Weiss, S.B. (1990) An environment-metapopulation approach to population viability analysis for a threatened invertebrate. *Conservation Biology*, **4**, 41–51.

Murphy, D.D. and Weiss, S.B. (1988) A long-term monitoring plan for a threatened butterfly. *Conservation Biology*, **2**, 367–374.

Murphy, D.D. and Wilcox, B.A. (1986) Butterfly diversity in natural habitat fragments: a test of the validity of vertebrate-based management, in *Wildlife 2000, Modelling Habitat Relationships of Terrestrial Vertebrates* (eds J.Verner, M.L. Morrison and C.J. Ralph), University of Wisconsin Press, Madison, pp. 287–292.

Myers, N. (1979) *The Sinking Ark,* Pergamon, New York.

Myers, N. (1988) Threatened biotas: 'Hotspots' in tropical forests. *Environmentalist*, **8**, 1–20.

Myers, N. (1990) The biodiversity challenge: expanded hot-spot analysis. *The Environmentalist*, **10**, 243–256.

Naess, A. (1989) *Ecology, Community and Lifestyle* (translated by D.Rothenberg), Cambridge University Press, Cambridge.

Naveh, Z. and Lieberman, A.S. (1990) *Landscape Ecology: Theory and Application,* Student Edition, Springer-Verlag, New York.

Nazzi, F., Paoletti, M.G. and Lorenzoni, G.G. (1989) Soil invertebrate dynamics of soybean agroecosystems encircled by hedgerows or not in

Friuli, Italy. First data. *Agriculture, Ecosystems and Environment,* **27,** 163–176.

Nepstad, D.C., Uhl, C. and Serrão, E.A.S. (1991) Recuperation of a degraded Amazonian landscape: forest recovery and agricultural restoration. *Ambio,* **20,** 248–255.

New, T.R. (1984) *Insect Conservation: An Australian Perspective,* Junk, Dordrecht.

New, T.R. (1987) *Butterfly Conservation,* Entomological Society of Victoria, Australia, Melbourne.

New, T.R. (1991) *Butterfly Conservation,* Oxford University Press, Melbourne.

New, T.R. and Collins, N.M. (1991) *Swallowtail Butterflies: An Action Plan for their Conservation,* International Union for Conservation of Nature and Natural Resources, Gland, Switzerland.

New, T.R. and Thornton, I.W.B. (1988) A pre-vegetation population of crickets subsisting on allochthonous aeolian debris on Anak Krakatau. *Philosophical Transactions of the Royal Society of London, B,* **322,** 481–485.

Nixon, K.C. and Wheeler, Q.D. (1990) An amplification of the phylogenetic species concept. *Cladistics,* **6,** 211–223.

Nixon, K.C. and Wheeler, Q.D. (1992) Measures of phylogenetic diversity, in *Evolution and Phylogeny* (eds M.J. Novacek, and Q.D. Wheeler), Columbia University Press, New York. pp. 216-234.

Norton, B.G. (1986) Epilogue, in *The Preservation of Species* (ed. B.G. Norton), Princeton University Press, Princeton, New Jersey, pp. 268–283.

Norton, B.G. (1987) *Why Preserve Natural Variety?,* Princeton University Press, Princeton, New Jersey.

Noss, R.F. (1987) From plant communities to landscapes in Conservation inventories: a look at The Nature Conservancy (USA). *Biological Conservation,* **41,** 11–37.

Noss, R.F. (1990) Indicators for monitoring biodiversity: A hierarchical approach. *Conservation Biology,* **4,** 355–364.

Noss, R.F. and Harris, L.D. (1986) Nodes, networks, and MUMS: Preserving diversity at all scales. *Environmental Management,* **10,** 299–309.

Oates, M.R. and Warren, M.S. (1990) *A Review of Butterfly Introductions in Britain and Ireland.* Report for the Joint Committee for the Conservation of British Insects funded by the World Wide Fund for Nature, 96 pp.

Odum, E.P., Connell, C.E. and Davenport, L.B. (1962) Population energy flow of three primary consumer components of old-field ecosystems. *Ecology,* **43,** 88–96.

Opdam, P. (1990) Dispersal in fragmented populations: the key to survival, in *Species Dispersal in Agricultural Habitats* (eds R.G.H. Bunce and D.C. Howard), Belhaven, London, pp. 3–17.

Opler, P.A. (1981) Management of prairie habitats for insect conservation. *Journal of the Natural Areas Association,* **1,** 3–6.

Opler, P.A. (1991) North American problems and perspectives, in *The Conservation of Insects and their Habitats* (eds N.M. Collins and J.A. Thomas), 15th Symposium of the Royal Entomological Society of London, Academic Press, London, pp. 9–32.

Ormerod, S.J., Weatherley, N.S. and Merrett, W.J. (1990) The influence of conifer plantations on the distribution of the Golden Ringed Dragonfly *Cordulegaster boltoni* (Odonata) in upland water. *Biological Conservation*, **53**, 241–151.

Owen D.F. and Owen J. (1974) Species diversity in temperate and tropical Ichneumonidae. *Nature*, **249**, 583–584.

Owen, D.F. and Owen J. (1990) Assessing insect species-richness at a single site. *Environmental Conservation*, **17**, 362–364.

Owen, J. and Gilbert, F.S. (1989) On the abundance of hoverflies (Syrphidae). *Oikos*, **55**, 183–193.

Owen, J. and Owen, D.F. (1975) Suburban gardens: England's most important nature reserves? *Environmental Conservation*, **2**, 53–59.

Parsons, M.J. (1984) The biology and conservation of *Ornithoptera alexandrae*, in *The Biology of Butterflies* (eds R.I. Vane-Wright and P.R. Ackery) Academic Press, London, pp. 327–331.

Parsons, P.A. (1982) Evolutionary ecology of Australian Drosophila: a species analysis. *Evolutionary Biology*, **14**, 297–347.

Parsons, P.A. (1989) Conservation and global warming: a problem in biological adaptation to stress. *Ambio*, **18**, 322–325.

Parsons, P.A. (1990) The metabolic cost of multiple environmental stresses: implications for climatic change and conservation. *Trends in Ecology and Evolution*, **5**, 315–317.

Parsons, P.A. (1991) Biodiversity conservation under global climatic change: the insect *Drosophila* as a biological indicator? *Global Ecology and Biogeography Letters*, **1**, 77–83.

Pearson, D.L. (1992) Tiger beetles as indicators for biodiversity patterns in Amazonia. *Research and Exploration*, **8**, 116–117.

Pearson, D.L. and Cassola, F. (1992) World-wide species richness patterns of tiger beetles (Coleoptera: Cicindelidae): indicator taxon for biodiversity and conservation studies. *Conservation Biology*, **6**, 376–391.

Pedgeley, D.E. (1982) *Windborne Pests and Diseases: Meteorology of Airborne Organisms,* Horwood, Chichester, UK.

Peterken, G.F. (1991) Ecological issues in the management of woodland nature reserves, in *The Scientific Management of Temperate Communities for Conservation* (eds I.F. Spellerberg, F.B. Goldsmith and M.G. Morris), Blackwell Scientific Publications, Oxford, pp. 245–272.

Peters, R.L. and Darling, J.D.S. (1985) The greenhouse effect and nature reserves. *BioScience*, **35**, 707–717.

Pianka, E.R. (1966) Latitudinal gradients in species diversity: a review of concepts. *The American Naturalist*, 100, 33–46.

Pielou, E.C. (1984) *The Interpretation of Ecological Data*, Wiley, New York.

Pimentel, D. (1975) Introduction, in *Insects, Science & Society* (ed. D. Pimentel), Academic Press, New York, pp. 1–10.

Pimentel, D. (1986) Agroecology and economics, in *Ecological Theory and Integrated Pest Management Practice* (ed. M. Kogan), Wiley, New York, pp. 299–319.

Pimentel, D., Culliney, T.W., Buttler, I.W., Reinemann, D.J. and Beckman, K.B. (1989) Low-input sustainable agriculture using ecological management practices. *Agriculture, Ecosystems and Environment*, 27, 3–24.

Pimm, S.L. (1986) Community stability and structure, in *Conservation Biology: The Science of Scarcity and Diversity* (ed. M.E. Soulé), Sinauer, Sunderland, Massachusetts, pp. 309–329.

Pimm, S.L. (1987) Determining the effects of introduced species. *Trends in Ecology and Evolution*, 2, 106–108.

Pimm, S.L. and Redfearn, A. (1988) The variability of population densities. *Nature*, 334, 613–614.

Pitelka, L.F. (1988) Evolutionary responses of plants to anthropogenic pollutants. *Trends in Ecology and Evolution*, 3, 233–236.

Platnick, N.I. (1991) Patterns of biodiversity: tropical v temperate. *Journal of Natural History*, 25, 1083–1088.

Plotkin, M.J. (1988) The outlook for new agricultural and industrial products from the tropics, in *Biodiversity* (ed. E.O. Wilson), National Academy Press, Washington, DC, pp. 106–116.

Pollard, E. (1981) Resource limited and equilibrium models of populations. *Oecologia (Berlin)*, 49, 377–378.

Pollard, E. (1982) Monitoring butterfly abundance in relation to the management of a nature reserve. *Biological Conservation*, 24, 317–328.

Pollard, E., Hooper, M. and Moore, N.W. (1974) *Hedges*, Collins, London.

Potter, V.R. (1988) *Global Bioethics*, Michigan State University Press, East Lansing, Michigan.

Prescott-Allen, R. and Prescott-Allen, C. (1982) *What's Wildlife Worth? Economic Contributions of Wild Plants and Animals to Developing Countries*, International Institute for Environment and Development, London.

Preston-Mafham, K. (1991) *Madagascar: A Natural History*, Facts on File, Oxford.

Price, P.W. (1984) *Insect Ecology*, 2nd edn, Wiley, New York.

Prinsloo, G.L. (1989) Insect identification services in South Africa. *Proceedings of the Seventh Entomological Congress organized by the Entomological Society of Southern Africa, Pietermaritzburg, 10–13 July 1989*, p. 107.

Pulliam, H.R. and Danielson, B.J. (1991) Sources, sinks and habitat selection: a landscape perspective on population dynamics. *The American Naturalist*, **137**, S50–S66.

Pye-Smith , C. and Rose, C. (1984) *Crisis and Conservation: Conflict in the British Countryside*, Penguin, Harmondsworth, England.

Pyle, R., Bentzien, M. and Opler, P. (1981) Insect Conservation. *Annual Review of Entomology*, **26**, 233–258.

Rabb, R.L., Stinner, R.E. and van den Bosch, R. (1976) Conservation and augmentation of natural enemies, in *Theory and Practice of Biological Control* (eds C.B. Huffaker and P.S. Messenger), Academic Press, New York, pp. 233–254.

Rabinowitz, D., Cairns, S. and Dillin, T. (1986) Seven forms of rarity and their frequency in the flora of the British Isles, in *Conservation Biology: The Science of Scarcity and Diversity* (ed. M.E. Soulé), Sinauer, Associates, Sunderland, Massachusetts, pp. 182–204.

Rackham, O. (1980) *Ancient Woodland*, Arnold, London.

Rackham, O. (1986) *The History of the Countryside,* Dent, London.

Ragge, D.R. (1974) Class Insecta, Order Orthoptera: Tettigonioidea, in *Taxonomy of the Hexapoda* (ed. W.G.H. Coaton), *Entomology Memoir No. 38.* Department of Agriculture of the Union of South Africa, Pretoria, pp. 37–38.

Rapoport, E.H. (1982) *Areography: Geographical Strategies of Species*, Pergamon, Oxford.

Reid, W.V. and Miller, K.R. (1989) *Keeping Options Alive: The Scientific Basis for Conserving Biodiversity*, World Resources Institute, Washington.

Reynolds, C.S. (1991) Lake communities: an approach to their management for conservation, in *The Scientific Management of Temperate Communities for Conservation* (eds I.F. Spellerberg, F.B. Goldsmith and M.G. Morris), Blackwell Scientific Publishers, Oxford, pp. 199–225.

Richards, O.W. and Davies, R.G. (1977) *Imm's General Textbook of Entomology,* 10th edn, Chapman & Hall, London.

Rohlf, D.J. (1991) Six biological reasons why the Endangered Species Act doesn't work – and what to do about it. *Conservation Biology*, **5**, 273–282.

Rolston III, H. (1986) *Philosophy Gone Wild,* Prometheus Books, Buffalo, New York.

Rolston III, H. (1988) *Environmental Ethics*, Temple University Press, Philadelphia.

Rosenberg, D.M., Danks, H.V. and Lehmkuhl, D.M (1986) Importance of insects in environmental impact assessment. *Environmental Management*, **10**, 773–783.

Rushton, S.P., Eyre, M.D. and Luff, M.L. (1990) The effects of scrub management on the ground beetles of oolitic limestone grassland at Castor

Hanglands National Nature Reserve, Cambridgeshire, UK. *Biological Conservation*, 51, 97–111.

Sailer, R.I. (1983) History of insect introductions, in *Exotic Plant Pests and North American Agriculture* (eds L. Wilson and L. Graham), Academic Press, New York, pp. 15–38.

Salo, J., Kalliola, R., Hakkinen, I., Makinen, Y., Niemela, P., Puhakka, M. and Coley, P.D. (1986), River dynamics and the diversity of Amazon Lowland forest. *Nature*, 322, 254–258.

Samways, M.J. (1976) Habitats and habits of *Platycleis* spp. (Orthoptera, Tettigoniidae) in southern France. *Journal of Natural History*, 10, 643–667.

Samways, M.J. (1977a) Bush cricket interspecific acoustic interactions in the field (Orthoptera: Tettigoniidae), *Journal of Natural History*, 11, 155–168.

Samways, M.J. (1977b) Effect of farming on population movements and acoustic behaviour of two bush crickets (Orthoptera: Tettigoniidae), *Bulletin of Entomological Research*, 67, 471–481.

Samways, M.J. (1979) Immigration, population growth and mortality of insects and mites on cassava (*Manihot esculenta*) in Brazil. *Bulletin of Entomological Research*, 69, 491–505.

Samways, M.J. (1981) *Biological Control of Pests and Weeds*, Edward Arnold, London.

Samways, M.J. (1983a) Community structure of ants (Hymenoptera : Formicidae) in a series of habitats associated with citrus. *Journal of Applied Ecology*, 20, 833–847.

Samways, M.J. (1983b) Inter-relationship between an entomogenous fungus and two ant–homopteran (Hymenoptera: Formicidae – Hemiptera: Pseudococcidae and Aphididae) mutualisms on guava trees. *Bulletin of Entomological Research*, 73, 321–331.

Samways, M.J. (1983c) Asymmetrical competition and amensalism through soil dumping by the ant, *Myrmicaria natalensis*. *Ecological Entomology*, 8, 191–194.

Samways, M.J. (1984a) Biology and economic value of the scale predator *Chilocorus nigritus* (F.) (Coccinellidae), *Biocontrol News and Information*, 5, 91–105.

Samways, M.J. (1984b) A practical comparison of diversity indices based on a series of small agricultural ant communities. *Phytophylactica*, 16, 275–278.

Samways, M.J. (1985) Relationship between red scale, *Aonidiella aurantii*(Maskell) (Hemiptera : Diaspididae), and its natural enemies in the upper and lower parts of citrus trees in South Africa. *Bulletin of Entomological Research*, 75, 379–393.

Samways, M.J. (1987) Weather and monitoring the abundance of the adult citrus psylla, *Trioza erytreae* (Del Guercio) (Hom., Triozidae), *Journal of Applied Entomology*, **103**, 502–508.

Samways, M.J. (1988a) A pictorial model of the impact of natural enemies on the population growth rate of the scale insect *Aonidiella aurantii. South African Journal of Science*, **84**, 270–272.

Samways, M.J. (1988b) Classical biological control and insect conservation: are they compatible? *Environmental Conservation*, **15**, 348–354.

Samways, M.J. (1989a) Insect conservation and landscape ecology: A case-history of bush crickets (Tettigoniidae) in southern France. *Environmental Conservation*, **16**, 217–226.

Samways, M.J. (1989b) Ecological landscaping at the National Botanic Gardens, Pietermaritzburg. *Veld & Flora*, **75**, 107–108.

Samways, M.J. (1989c) Insect conservation and the disturbance landscape. *Agriculture, Ecosystems and Environment*, **27**, 183–194.

Samways, M.J. (1989d) Climate diagrams and biological control: an example from the areography of the ladybird *Chilocorus nigritus* Fabricius, 1798) (Insecta, Coleoptera, Coccinellidae), *Journal of Biogeography*, **16**, 345–351.

Samways, M.J. (1989e) Taxon turnover in Odonata across a 3000m altitudinal gradient in southern Africa. *Odonatologica*, **18**, 263–274.

Samways, M.J. (1989f) Farm dams as nature reserves for dragonflies (Odonata) at various altitudes in the Natal Drakensberg Mountains, South Africa. *Biological Conservation*, **48**, 181–187.

Samways, M.J. (1990a) Species temporal variability: epigaeic ant assemblages and management for abundance and scarcity. *Oecologia*, **84**, 482–490.

Samways, M.J. (1990b) Landforms and winter habitat refugia in the conservation of montane grasshoppers in southern Africa. *Conservation Biology*, **4**, 375–382.

Samways, M.J. (1990c) Ant assemblage structure and ecological management in citrus and subtropical fruit orchards in southern Africa, in *Applied Myrmecology: A World Perspective* (eds R.K. Van der Meer, K. Jaffe and A. Cedano), Westview Press, Boulder, Colorado, pp. 570–587.

Samways, M.J. (1992a) Dragonfly conservation in South Africa: a biogeographical perspective. *Odonatologica* **21**, 165–180.

Samways, M.J. (1992b) Some comparative insect conservation issues of north temperate, tropical, and south temperate landscapes. *Agriculture, Ecosystems and Environment* **40**, 137–154.

Samways, M.J. (1993) Threatened Lycaenidae of South Africa, in *Threatened Lycaenidae of the World* (ed. T.R. New) IUCN, Gland.

Samways, M.J. and Grech, N.M. (1986) Assessment of the fungus *Cladosporium oxysporum* (Berk. and Curt.) as a potential biocontrol agent against certain Homoptera. *Agriculture, Ecosystems and Environment*, **15**, 231–239.

Samways, M.J. and Harz, K. (1982) Biogeography of intraspecific morphological variation in the bush crickets *Decticus verrucivorus* (L.) and *D. albifrons* (F.) (Orthoptera: Tettigoniidae), *Journal of Biogeography* **9**, 243–254.

Samways, M.J. and Manicom, B.Q. (1983) Immigration, frequency distributions and dispersion patterns of the psyllid *Trioza erytreae* (Del Guercio) in a citrus orchard. *Journal of Applied Ecology*, **20**, 463–472.

Samways, M.J. and Moore, S.D. (1991) Influence of exotic conifer patches on grasshopper (Orthoptera) assemblages in a grassland matrix at a recreational resort, Natal, South Africa. *Biological Conservation*, **57**, 205–219.

Samways, M.J., Nel, M. and Prins, A.J. (1982) Ants (Hymenoptera: Formicidae) foraging in citrus trees and attending honeydew-producing Homoptera. *Phytophylactica*, **14**, 155–157.

Sanford, R.L., Saldarriaga, J., Clark, K.E., Uhl, C. and Herrera, R. (1985) Amazon rain-forest fires. *Science (NY)*, **227**, 53–55.

Sant, G.J. and New, T.R. (1988) The biology and conservation of *Hemiphlebia mirabilis* Selys (Odonata, Hemiphlebiidae) in southern Victoria. Arthur Rylah Institute for Environmental Research, Technical Report Series No. 82., Melbourne, pp. 1–35.

Saunders, D.A. and Hobbs, R.J. (eds) (1991) *Nature Conservation 2: The Role of Corridors*, Surrey Beatty, Chipping Norton, New South Wales, Australia.

Saunders, D.A., Arnold, G.W., Burbridge, A.A. and Hopkins, A.J.M. (eds) (1987) *Nature Conservation: The Role of Remnants of Native Vegetation*, Surrey Beatty, Chipping Norton, New South Wales, Australia.

Schatz, G.E. (1992) The race between deep flowers and long tongues. *Wings*, Summer 1992, 19–21.

Schitz, A. (1989) Conserving biological diversity: Who is responsible? *Ambio*, **18**, 454–457.

Schneider, S.H. (1989) The changing climate. *Scientific American*, **261**, 38–46.

Schneider, S.H. and Londer, R. (1984) *The Ecoevolution of Climate and Life*, Sierra Club Books, San Francisco.

Schoener, T.W. (1986) Patterns in terrestrial vertebrate versus arthropod communities: Do systematic differences in regularity exist? in *Community Ecology*, (eds J. Diamond and T.J. Case), Harper & Row, New York.

Schoener, T.W. (1987) The geographical distribution of rarity. *Oecologia (Berlin)*, **74**, 161–173.

Schoener, T.W. (1990) The geographical distribution of rarity: misinterpretation of atlas methods affects some empirical conclusions. *Oecologica*, **82**, 567–568.

Schoener, T.W. and Janzen, D.H. (1968) Notes on environmental determinants of tropical versus temperate insect size patterns. *The American Naturalist*, **102**, 207–224.

Scholes, R.J. (1990) Change in nature and the nature of change: interactions between terrestrial ecosystems and the atmosphere. *South African Journal of Science*, **86**, 350–354.

Schonewald-Cox C. (1988) Boundaries in the protection of nature reserves. *BioScience*, **38**, 480–486.

Schowalter, T.D. (1985) Adaptions of insects to disturbance, in *The Ecology of Natural Disturbance and Patch Dynamics* (eds S.T.A. Pickett and P.S. White), Academic Press, San Diego, pp. 235–252.

Schowalter, T.D., Coulson, R.N. and Crossley, D.A. Jr, (1981) Role of southern pine beetle and fire in maintenance of structure and function of the south eastern coniferous forests. *Environmental Entomology*, **10**, 821–825.

Schumacher, E.F. (1973) *Small is Beautiful*, Abacus, London.

Schwerdtfeger, F. (1941) Über die Ursachen des Massenwechsels der Insekten. *Zeitschrift für Angewandte Entomologie*, **28**, 254–303.

Scourfield, M.W.J., Bodecker, G., Barker, M.D., Diab, R.D. and Salter, L.F. (1990) Ozone: the South African connection. *South African Journal of Science*, **86**, 279–281.

Shirt, D.B. (ed.) (1987) *British Red Data Books: 2. Insects,* Nature Conservancy Council, Peterborough, London.

Shure, D.J. and Phillips, D.L. (1991) Patch size of forest openings and arthropod populations. *Oecologia*, **86**, 325–334.

Silvestri, F. (1913) Descrizione di un nuovo ordine di insetti. *Bollettino del Laboratoria di Zoologia, Portici*, **7**, 193–209.

Simpson, G.G. (1952) How many species? *Evolution*, **6**, p.432.

Sioli, H. (1986) Tropical continental aquatic habitats, in *Conservation Biology: The Science of Scarcity and Diversity* (ed. M.E. Soulé), Sinauer, Sunderland, Massachusetts, pp. 383–393.

Smith, R.H. and Holloway, G.J. (1989) Population genetics and insect introductions. *The Entomologist*, **108**, 14–27.

Sotherton, N.W., Boatman, N.D. and Rands, M.R.W. (1989) The 'Conservation Headland' experiment in cereal ecosystems. *The Entomologist*, **108**, 135–143.

Soulé, M.E. (1986) (ed.) *Conservation Biology: The Science of Scarcity & Diversity*, Sinauer, Sunderland, Massachusetts.

Soulé, M.E. (1989) Conservation biology in the twenty-first century: summary and outlook, in *Conservation for the Twenty-first Century* (ed. D. Western and M.C. Pearl), Oxford University Press, New York, pp. 297–303.

Soulé, M.E. and Gilpin, M.E. (1991) The theory of wildlife corridor capability, in *Nature Conservation 2. The Role of Corridors*, (eds D.S. Saunders and R.J. Hobbs), Surrey Beatty, Chipping Norton, New South Wales, Australia, pp. 3–8.

Soulé, M.E. and Wilcox, B.A. (eds) (1980) *Conservation Biology: An Evolutionary – Ecological Perspective*, Sinauer, Sunderland, Massachusetts.

South, R. (1906) *The Butterflies of the British Isles*, Warne, London.

Southwood, T.R.E. (1961) The number of species of insect associated with various trees. *Journal of Animal Ecology*, 30, 1–8.

Southwood, T.R.E.(1973) The insect/plant relationship – an evolutionary perspective. *Symposium of the Royal Entomological Society, London*, 6, 3–30.

Southwood, T.R.E. (1978) *Ecological Methods: With Particular Reference to the Study of Insect Populations*, 2nd edn, Chapman & Hall, London.

Southwood, T.R.E. (1988) Tactics, strategies and templets. *Oikos*, 52, 3–18.

Southwood, T.R.E., Brown, V.K., and Reader, P.M. (1979) The relationships of plant and insect diversities in succession. *Biological Journal of the Linnaean Society*, 12, 327–348.

Spellerberg, I.F. (1981) *Ecological Evaluation for Conservation*, Edward Arnold, London.

Spellerberg, I.F. (1991) Biogeographical basis of Conservation, in *The Scientific Management of Temperate Communities for Conservation* (eds I.F. Spellerberg, F.B. Goldsmith and M.G. Morris), Blackwell Scientific Publications, Oxford, pp. 293–322.

Stamps, J.A., Buechner, M. and Krishnan, V.V. (1987) The effects of edge permeability and habitat geometry on emigration from patches of habitat. *The American Naturalist*, 129, 533–552.

Steele, R.C. and Welch, R.C. (eds) (1973) *Monks Wood, A Nature Reserve Record*, Nature Conservancy, Monks Wood Experimental Station, England.

Stevens, G.C. (1989) The latitudinal gradient in geographic range: how so many species co-exist in the tropics. *The American Naturalist*, 133, 240–256.

Stewart, L.M.D., Hirst, M., Ferber, M.L., Merryweather, A.T., Cayley, P.J. and Possee, R.D. (1991) Construction of an improved baculovirus insecticide containing an insect-specific toxin gene. *Nature*, 352, 85–88.

Stinner, D.H., Paoletti, M.G. and Stinner, B.R. (1989) In search of traditional farm wisdom for more sustainable agriculture: a study of Amish farming and society. *Agriculture, Ecosystems and Environment*, 27, 77–90.

Stinner, R.E., Rabb, R.L. and Bradley, J.R. Jr, (1974) Population dynamics of *Heliothis zea* (Boddie) and *H. virescens* (F.) in North Carolina: a simulation model. *Environmental Entomology*, 3, 163–168.

Stork, N.E. (1988) Insect diversity: facts, fiction and speculation. *Biological Journal of the Linnean Society*, 35, 321–337.

Stork, N.E. (1991) The comparison of the arthropod fauna of Bornean lowland rain forest trees. *Journal of Tropical Ecology*, 7, 161–180.

Stout, J. and Vandermeer, J. (1975) Comparison of species richness for stream inhabiting insects in tropical and mid-latitude streams. *The American Naturalist*, **109**, 263–280.

Strong, D.R., Lawton, J.H. and Southwood, Sir R. (1984) *Insects on Plants: Community Patterns and Mechanisms*, Blackwell Scientific Publications, Oxford.

Sugimura, M. (1989) *Dragonfly Kingdom*, World Wide Fund for Nature, Japan. (In Japanese)

Sutherst, R.W. (1991) Pest risk analysis and the greenhouse effect. *Review of Agricultural Entomology*, **79**, 1177-1187.

Sutton, S.L. and Collins, N.M. (1991) Insects and tropical forest conservation, in *The Conservation of Insects and their Habitats* (eds N.M. Collins and J.A. Thomas), 15th Symposium of the Royal Entomological Society of London, Academic Press, London, pp. 405–424.

van Swaay, C.A.M. (1990) An assessment of the changes of butterfly abundance in The Netherlands during the 20th century. *Biological Conservation*, **52**, 287–302.

Swanson, F.J., Kratz, T.K., Caine, N. and Woodmansee, R.G. (1988) Landform effects and ecosystem patterns and processes. *BioScience*, **38**, 92–98.

Swinton, A.H. (*c.* 1880) *Insect Variety: Its Propagation and Distribution*, Cassell, Petter & Galpin, London.

Tainton, N.M. and Mentis, M.T. (1984) Fire in grassland, in *Ecological Effects of Fire in South African Ecosystems* (eds P. de V. Booysen and N.M. Tainton), Springer-Verlag, Berlin, pp. 115–147.

Taylor, J.C. (1979) *Sensoriaphis* in Western Australia. *Bulletin of the Royal Entomological Society, London*, **3**, p.106.

Taylor, L.R., French, R.A. and Woiwood, I.P. (1978) The Rothamsted insect survey and the urbanization of land in Great Britain, in *Perspectives in Urban Entomology* (eds G.W. Frankie and C.S. Koehler), Academic Press, London, pp. 31–65.

Taylor, P.W. (1986) *Respect for Nature: A Theory of Environmental Ethics*, Princeton University Press, Princeton, New Jersey.

Thomas, C.D. (1991) Habitat use and geographic ranges of butterflies from the wet lowlands of Costa Rica. *Biological Conservation*, **55**, 269–281.

Thomas, J.A. (1980) Why did the Large Blue become extinct in Britain? *Oryx*, **15**, 243–247.

Thomas, J.A. (1983a) The ecology and conservation of *Lysandra bellargus* in Britain. *Journal of Applied Ecology*, **20**, 59–83.

Thomas, J.A. (1983b) *Maculinea arion*, in *British Red Data Books. 2. Insects.* (ed. D.B. Shirt), Nature Conservancy Council, Peterborough, pp. 79–80.

Thomas, J.A. (1984) The conservation of butterflies in temperate countries: past efforts and lessons for the future, in *The Biology of Butterflies* (eds R.I. Vane-Wright and P.R. Ackery), Academic Press, London, pp. 333–353.

Thomas, J.A. (1989) Ecological lessons from the re-introduction of Lepidoptera. *The Entomologist,* 56–68.

Thomas, J.A. (1991) Rare species conservation: case studies of European butterflies, in *The Scientific Management of Temperate Communities for Conservation* (eds I.F. Spellerberg, F.B. Goldsmith and M.G. Morris), Blackwell Scientific Publications, Oxford, pp. 149–197.

Thomas, J.A., Thomas, C.D., Simcox, D.J. and Clarke, R.J. (1986) Ecology and declining status of the silver-spotted skipper *(Hesperia comma)* in Britain. *Journal of Applied Ecology,* 23, 365–380.

Tilman, D. (1982) *Resource Competition and Community Structure,* Princeton University Press, Princeton, New Jersey.

Todd, J. (1988) Restoring biodiversity: the search for a social and economic context, in *Biodiversity* (ed. E.O.Wilson), National Academy Press, Washington DC, pp. 344–352.

van Tol, J. and Verdonk, M.J. (1988) *The Protection of Dragonflies (Odonata) and their Biotopes.* European Committee for the Conservation of Nature and Natural Resources, Strasbourg.

Tomalski, M.D. and Miller, L.K. (1991) Insect paralysis by baculovirus-mediated expression of a mite neurotoxin gene. *Nature,* 352, 82–85.

Turin, H. and den Boer, P.J. (1988) Changes in the distribution of carabid beetles in The Netherlands since 1880: II. Isolation of habitats and long-term time trends in the occurrence of carabid species with different powers of dispersal (Coleoptera, Carabidae). *Biological Conservation,* 44, 179–200.

Turner, M.G. and Gardner, R.H. (eds) (1991) *Quantitative Methods in Landscape Ecology: The Analysis and Interpretation of Landscape Heterogeneity,* Springer-Verlag, New York.

Turner, M.G., Gardner, R.H., Dale, V.H. and O'Neill, R.V. (1989) Predicting the spread of disturbance across heterogenous landscapes. *Oikos,* 55, 121–129.

Turrill, W.B. (1948) *British Plant Life,* Collins, London.

Uetz, G.W. (1991) Habitat structure and spider foraging, in *Habitat Structure: The Physical Arrangement of Objects in Space* (eds S.S. Bell, E.D. McCoy and H.R. Mushinsky), Chapman & Hall, London, pp. 325–348.

Ugalde, A.F. (1989) An optional parks system. in *Conservation for the Twenty-first Century* (eds D. Western and M.C. Pearl), Oxford University Press, New York, pp. 145–149.

Usher, M.B. (ed.) (1986a) *Wildlife Conservation Evaluation,* Chapman & Hall, London.

Usher, M.B. (1986b) Wildlife conservation evaluation: attributes, criteria and values, in *Wildlife Conservation Evaluation* (ed. M.B. Usher), Chapman & Hall, London, pp. 3–44.

Usher, M.B. (1992) Management and diversity of arthropods in *Calluna* heathland. *Biodiversity and Conservation,* 1, 63-79.

Usher, M.B. and Jefferson, R.G. (1991) Creating new and successional habitats for arthropods, in *The Conservation of Insects and their Habitats* (eds N.M. Collins and J.A. Thomas), Academic Press, London, pp. 263–291.

Vane-Wright, R.I., Humphries, C.J. and Williams, P.H. (1991) What to protect? – Systematics and the agony of choice. *Biological Conservation,* 55, 235–254.

Varley, G.C., Gradwell, G.R. and Hassell, M.P. (1973) *Insect Population Ecology,* Blackwell Scientific Publications, Oxford.

Waage, J.K. and Greathead, D.J. (1988) Biological control: challenges and opportunities. *Philosophical Transactions of the Royal Society of London, B,* 318, 118–128.

Wallace, J.B. (1990) Recovery of lotic macroinvertebrate communities from disturbance. *Environmental Management,* 14, 605–620.

Walter, H. (1985) *Vegetation of the Earth,* 3rd edn, Springer-Verlag, Berlin.

Walter, H. and Leith, H. (1960–67) *Klimadiagramm-welt Atlas,* Gustav-Fischer, Stuttgart.

Warren, M.S. (1984) The biology and status of the wood white butterfly, *Leptidea sinapis* (Lepidoptera: Pieridae) in the British Isles. *Entomologists' Gazette,* 35, 207–223.

Warren, M.S. (1985) Influence of shade on butterfly numbers in woodland rides with special reference to the wood white *Leptida sinapis. Biological Conservation,* 33, 147–164.

Warren, M.S. (1987a) The ecology and conservation of the heath fritillary butterfly *Mellicta athalia II.* Adult population structure and mobility. *Journal of Applied Ecology,* 24, 483–498.

Warren, M.S. (1987b) The ecology and conservation of the heath fritillary butterfly, *Mellicta athalia III.* Population dynamics and the effect of habitat management. *Journal of Applied Ecology,* 24, 499–514.

Warren, M.S. and Key, R.S. (1991) Woodlands: past, present and potential for insects, in *The Conservation of Insects and their Habitats* (eds N.M. Collins and J.A. Thomas), 15th Symposium of the Royal Entomological Society of London, Academic Press, London, pp. 155–211.

Warren, S.D., Scifres, C.J. and Teel, P.D. (1987) Response of grassland arthropods to burning: a review. *Agriculture, Ecosystems and Environment,* 19, 105–130.

Watson, J.A.L., Arthington, A.H. and Conrick, D.L. (1982) Effect of sewage effluent on dragonflies (Odonata) of Bulimba Creek, Brisbane. *Australian Journal of Marine and Freshwater Research,* 33, 517–528.

Webb, N.R. (1989) Studies on the invertebrate fauna of fragmented heathland in Dorset, UK, and the implications for conservation. *Biological Conservation,* 47, 153–165.

Weiss, S.B., Rich, P.M., Murphy, D.D., Calvert, W.H. and Ehrlich, P.R. (1991) Forest canopy structure at overwintering Monarch Butterfly sites: measurements with hemispherical photography. *Conservation Biology*, 5, 165–175.

Welch, R.C. (1990) Dispersal of invertebrates in the agricultural environment, in *Species Dispersal in Agricultural Habitats* (eds R.G.H. Bunce and D.C. Howard), Belhaven, London, pp. 203–218.

Wells, G. (1989) Observing earth's environment from space, in *The Fragile Environment* (eds L. Friday and R. Laskey), Cambridge University Press, Cambridge, pp. 148–192.

Wells, S.M., Pyle, R.M. and Collins, N.M. (1983) *The IUCN Invertebrate Red Data Book*, IUCN, Gland, Switzerland.

Wells, T.C.E. (1989) The re-creation of grassland habitats. *The Entomologist*, 108, 97–108.

Wheeler, Q.D. (1990) Insect diversity and cladistic constraints. *Annals of the Entomological Society of America*, 83, 1031–1047.

White, E.B., DeBach, P. and Garber, M.J. (1970) Artificial selections for genetic adaptation to temperature extremes in *Aphytis lingnanensis* Compere (Hymenoptera: Aphelinidae), *Hilgardia*, 40, 161–192.

White, P.S. and Pickett, S.T.A. (1985) Natural disturbance and patch dynamics: an introduction, in *The Ecology of Natural Disturbance and Patch Dynamics* (eds S.T.A. Pickett and P.S. White), Academic Press, New York, pp. 3–13.

Whittaker, R.H. (1975) *Communities and Ecosystems*, 2nd edn, Macmillan, New York.

Widgery, J.P. (1978) Roesel's bush-cricket *Metrioptera roeselii* in Regent's Park. *London Naturalist*, 57, 57–58.

Wigglesworth, V.B. (1976) *Insects and the Life of Man*, Chapman & Hall, London.

Wilcove, D.S., McLellan, C.H. and Dobson, A.P. (1986) Habitat fragmentation in the temperate zone, in *Conservation Biology: The Science of Scarcity and Diversity* (ed. M.E. Soulé), Sinauer Associates, Sunderland, Massachusetts, pp. 237–256.

Wilcox, B.A. and Murphy, D.D. (1985) Conservation strategy: the effects of fragmentation on extinction. *The American Naturalist*, 125, 879–887.

Williams, P.H., Humphries, C.J. and Vane-Wright, R.I. (1991) Measuring biodiversity: taxonomic relatedness for conservation priorities. *Australian Systematic Botany*, 4, 665–679.

Williams-Linera, G. (1990) Vegetation structure and environmental conditions of forest edges in Panama. *Journal of Applied Ecology*, 78, 356–373.

Williamson, M. (1972) *The Analysis of Biological Populations*, Edward Arnold, London.

Williamson, M. (1984) The measurement of population variability. *Ecological Entomology*, 9, 239–241.

Williamson, M. (1991) Biocontrol risks. *Nature*, 353, 394.

Williamson, M.H. and Lawton, J.H. (1991) Fractal geometry of ecological habitats, in *Habitat Structure: The Physical Arrangement of Objects in Space* (eds S.S. Bell, E.D. McCoy and H.R. Mushinsky), Chapman & Hall, London, pp. 69–86.

Wilson, E.O. (1985) The biological diversity crisis. *BioScience*, 35, 700–706.

Wilson, E.O. (1987) The little things that run the world (the importance and conservation of invertebrates), *Conservation Biology*, 1, 344–346.

Wilson E.O. (ed.) (1988) *Biodiversity*, National Academy Press, Washington DC.

Wilson, E.O. (1991) Ants. *Wings*, No. 16, 3–13.

Wilson, E.O. and Willis, E.O. (1975) Applied biogeography, in *Ecology and Evolution of Communities* (eds M.L. Cody and J.M. Diamond), Belknap Press of Harvard University, Cambridge, Massachusetts, pp. 523–534.

Wilson, F. (1960) A review of the biological control of insects and weeds in Australia and Australian New Guinea. *Technical Communication No. 1*, Commonwealth Institute of Biological Control, Commonwealth Agricultural Bureaux, Farnham Royal, England.

Wolda, H. (1987) Altitude, habitat and tropical insect diversity. *Biological Journal of the Linnean Society*, 30, 313–323.

Wolda, H. (1988) Insect seasonality: Why? *Annual Review of Ecology and Systematics*, 19, 1–18.

Wolda, H. (1992) Trends in abundance of tropical forest insects. *Oecologia* 89, 47–52.

Wood, J.G. (1863) *The Illustrated Natural History*, Routledge, Warne & Routledge, London.

Wood, P.A. and Samways, M.J. (1991) Landscape element pattern and continuity of butterfly flight paths in an ecologically landscaped botanic garden, Natal, South Africa. *Biological Conservation*, 58, 149–166.

Woodward, F.I. (1990) Global change: translating plant ecophysiological responses to ecosystems. *Trends in Ecology and Evolution*, 5, 308–311.

Woodwell, G.M. and Ramakrishna, K. (1989) The warming of the earth: perspectives and solutions in the third world. *Environmental Conservation*, 16, 289–291.

Wright, J.F., Moss, D., Armitage, P.D. and Furse, M.T. (1984) A preliminary classification of running-water sites in Great Britain based on macro-invertebrate species and the prediction of community type using environmental data. *Freshwater Biology*, 14, 221–256.

Yahner, R.H. (1988) Changes in wildlife communities near edges. *Conservation Biology*, 2, 333–339.

Yonzon, P., Jones, R. and Fox, J. (1991) Geographic information systems for assessing habitat and estimating population of red pandas in Langtang National Park, Nepal. *Ambio,* **20,** 285–288.

Zanaboni, A. and Lorenzoni, G.G. (1989) The importance of hedges and relict vegetation in agroecosystems and environment reconstruction. *Agriculture, Ecosystems and Environment,* **27,** 155–161.

Zgomba, M., Petrovic, D. and Srdic, Z. (1986) Mosquito larvicide impact on mayflies (Ephemeroptera) and dragonflies (Odonata) in aquatic biotopes. *Odonatologica,* **16,** 221–222.

Zimmerman, E.C. (1958) *Insects of Hawaii: Vol.7 (Macrolepidoptera),* University of Hawaii Press, Honolulu.

Zimmerman, H.G. and Moran, V.C. (1991) Biological control of prickly pear, *Opuntia ficus-indica* (Cactaceae), in South Africa. *Agriculture, Ecosystems and Environment,* **37,** 29–35.

Index

Page numbers in **bold** refer to figures